UWB Technology - New Insights and Developments

Edited by Rafael Vargas-Bernal

Published in London, United Kingdom

UWB Technology - New Insights and Developments
http://dx.doi.org/10.5772/intechopen.104029
Edited by Rafael Vargas-Bernal

Contributors
Ahmad - Yarahmadi, Abhishek Joshi, Rahul Singhal, Jerome Henry, Oliver Michler, Paul Schwarzbach, Jonas Ninnemann, Rafael Vargas-Bernal

First published in London, United Kingdom, 2023 by IntechOpen
IntechOpen is the global imprint of INTECHOPEN LIMITED, registered in England and Wales, registration number: 11086078, 5 Princes Gate Court, London, SW7 2QJ, United Kingdom

British Library Cataloguing-in-Publication Data
A catalogue record for this book is available from the British Library

Additional hard and PDF copies can be obtained from orders@intechopen.com

UWB Technology - New Insights and Developments
Edited by Rafael Vargas-Bernal
p. cm.
Print ISBN 978-1-83768-547-9
Online ISBN 978-1-83768-548-6
eBook (PDF) ISBN 978-1-83768-549-3

Meet the editor

Rafael Vargas-Bernal obtained his degree in electronics and communications engineering from the University of Guanajuato in 1995, and his master's and Doctor of Science degrees in electronics from the National Institute of Astrophysics, Optics and Electronics (INAOE) in 1997 and 2000, respectively. Since 2000, he has held the position of full professor at the University of Guanajuato, the Instituto Tecnológico Superior de Irapuato, and Universidad de la Salle Bajío. He is a reviewer for *ACS, RSC, Elsevier, IEEE, MDPI, Hindawi,* and *Wiley* journals. He has published 17 articles in indexed journals and 37 book chapters. He has published one book for IntechOpen on hybrid nanomaterials. His areas of interest are wireless communications, nanomaterials, and biomedical devices.

Contents

Preface

The urgent need to develop wireless broadband communication systems is driving research into both devices and applications for delivering large amounts of information at high data rates. One of the serious candidate technologies to achieve these capabilities is ultra-wideband (UWB). This technology uses very narrow pulses with nanosecond periods and can achieve wide bandwidths over short distances using low power densities. Applications such as real-time location systems for industrial and medical applications are made possible by UWB technology. With the need for more secure and cost-effective options than Bluetooth or Wi-Fi, there is growing research interest in UWB technology across the world, and smart devices based on UWB technology will be in widespread use in the coming decades.

This book presents work by different researchers around the world on both devices and applications based on UWB technology. The contributions are arranged in two sections covering UWB technology-based devices and applications, respectively.

Chapter 1 introduces different alternatives that UWB technology is implementing to improve its device design capabilities, taking advantage of materials science to develop more accurate radio frequency location systems. Potential devices include transmitters, antennas, filters, and resonators for applications in emerging UWB systems. A polygonal patch antenna proposed in Chapter 2 is one of the novel devices that can be implemented using UWB technology. This type of antenna can operate in wide bands, reducing the effect of the ground plane through computer simulation of electrical models using a hexagonal geometry with a high gain. The authors exploit a reflector based on an artificial magnetic conductor that offers promising capabilities for the development of UWB technology.

Wideband real-time delay cells for wideband multiple antennas for timed array receivers are proposed in Chapter 3. To optimize the performance required for UWB technology, a compact silicon-based integrated circuit design is suggested for the antennas, based on transmission lines and ladder networks operating at GHz. Active all-pass filters as TTD cells are presented for this purpose.

Chapter 4 examines the SS-TWR (single-sided two-way ranging) and DS-TWR (double-sided two-way ranging) techniques that exploit the 802.15.4a and 802.15.4z standards to develop navigation and active tracking systems. With bidirectional exchanges between the client and the anchor, it is possible to determine the distance between the transmitters and receivers to optimize signal arrival times to achieve wireless communication using UWB technology.

Another innovative application of UWB technology is the development of radio impulse sensing presented in Chapter 5. This type of sensing exploits the integration of location and radar functionalities to transmit signals through hardware with high resolution and multiple communication channels, enabling mobility and transportation applications.

This book will be a valuable source for undergraduate and graduate students, as well as for experts and researchers around UWB technology. I appreciate all the time and effort contributed by the research authors. I am also grateful to the staff at IntechOpen, and especially to Ms. Ana Cink for her valuable support during the book editing process. Finally, I wish to thank my wife, my son, my parents, brothers, and sisters for their help and support.

Dr. Rafael Vargas-Bernal
Center for Research and Development in Micro
and Nano Technologies (CIDEMYNT),
Higher Technological Institute of Irapuato,
Irapuato, Guanajuato, México

Section 1

UWB Technology-Based Devices

Chapter 1

Introductory Chapter: Novel Developments in UWB Technology

Rafael Vargas-Bernal

1. Introduction

Like any technology that evolves, UWB systems must reduce their complexity, power consumption, sustainability, and the possibility of reconfiguration, to achieve the highest performance compared to other competitors for implementations of high-accuracy localization systems [1–8]. Despite the advances achieved so far, UWB systems face challenges such as mitigating errors from non-line-of-sight paths and jamming signal interference in dense environments, especially in extreme conditions [1]. In addition, in the pursuit of an upgrade, UWB systems must integrate machine learning capabilities as well as sensor data fusion. The main disadvantages of UWB technology are its high cost and increased power consumption. Until now, it has been established that UWB technology presents interference with the radio frequency systems found in its surroundings and vice versa. The data capacity of UWB systems is restricted because short-duration pulse coding implies longer information synchronization times. The purpose of this chapter is to present the different alternatives that UWB technology is investigating to improve its properties using integrated circuit design as well as materials science and engineering to establish itself as an emerging strategy for its application in high-accuracy radio frequency location systems.

The remainder of the chapter is divided as follows: Section 2 introduces the basic concepts associated with UWB technology. Novel approaches for the use of UWB technology are summarized in Section 3. Some new developments in UWB systems are discussed in Section 4. Finally, the conclusions of the chapter are provided.

2. Basic concepts

The Federal Communications Commission (FCC) has established that for medical applications, Ultra-Wide Band (UWB) is set in the range of 3.1–10.6 GHz [1–8]. UWB systems represent a technological alternative for the development of applications such as the Internet of Things (IoT), energy harvesting, biomedical, and wireless communication systems [5]. Among the significant advantages that UWB systems have are their 7.5 GHz bandwidth, high data rate, reduced complexity, low power consumption, narrowband interference attenuation, as well as multiple transceiver architectures for different ranges. Also, the bandwidth is greater than 500 MHz or has a value of 20 percent of the center frequency. The power spectral density has a value of less than −41.3 dBm/MHz across the frequency band. For indoor positioning applications, UWB systems have short message lengths, high data rates, and bandwidths, low transmission powers, as well as high penetration capabilities. UWB signals have multiple frequency components to penetrate obstacles in the signal transmission paths [1]. The spectrum of UWB systems was allocated as free since 2002 for commercial use. Thanks to the low power spectral density, UWB systems do not interfere with other radio frequency signals. Furthermore, high accuracy and good multipath performance are possible due to the short pulse duration that UWB systems have. The specific applications that UWB systems have at the industrial level are smart logistics, the smart city, the smart factory, vehicle tracking, waste management, and robot positioning. The use of UWB technology in internal logistics allows for combining precision, reliability, and scalability in the tracking of goods and people, as well as in the automated control of vehicles. Smart logistics for real-time applications is necessary to track resources, materials, and employees at the same time to plan strategies to optimize the use of UWB technology.

So far six different categories have been developed for UWB signals [1, 8]. In an ultra-wideband system using radio pulses with pulses on the order of nanoseconds, it presents a low-duty cycle for transferring information by varying the phase, pulse shape, duration, and amplitude of the radio signal used [1]. This technology known as IR-UWB is governed by the IEEE 802.15.4z standard. Pseudorandom coding is used for ultra-wideband (UWB) technology through direct sequences called (DS-UWB) which is produced by amplitude modulation of a set of short pulses. Through Orthogonal Frequency Division Multiplexing (OFDM), it is possible to take advantage of the full bandwidth by dividing it into multiple frequency sub-bands using Quadrature Phase Shift Keying (QPSK) modulation to produce multi-bandwidth UWB systems (MB-UWB). The use of frequency-hopping ultra-bandwidth (FH-UWB) systems using variant frequency carriers can be used for periodic narrowband transmission. When frequency hopping involves selection through codes using discrete steps until the bandwidth is reached, it is called the Stepped Frequency Hopping UWB (SFH-UWB) system. The latter category involves the carrier frequency variation being generated by a voltage-controlled oscillator via continuously variable speed which is known as a Swept Frequency UWB (SF-UWB) system.

Typical UWB-based positioning systems include fixed sensors known as anchors, moving targets known as tags, a location server, and a system interface [1]. The location server has the function of storing and processing the data provided by the sensors, while the system interface, commonly a smartphone, computer, or tablet, allows viewing the positioning results. For a two-dimensional positioning system, at least three anchors are required to operate. In the case of a three-dimensional system, it is necessary to use at least four anchors. In addition, more complex systems involving the integration of environments such as the Internet of Things (IoT) or multi-sensor technologies lead to the use of more sophisticated and intelligent user interfaces, network gateways, and navigation frameworks.

3. Novel approaches for the use of ultra-wideband (UWB) technology

Wearable tracking systems are and will be important for monitoring the physical activity of high-performance athletes through ultra-wideband (UWB) positioning sensors to determine performance parameters such as speed, distance covered, acceleration, and change of direction of travel [3]. Until now, UWB systems achieve a positioning accuracy of up to 10 cm, because this decreases by half to one meter in a three-dimensional location. This positioning tracking is also required to determine the round-trip time of robots indoors or in real-time bus parking and tracking or patient location tracking inside and outside hospitals. The design of UWB devices must address the feasibility of tuning using alternative configurations, different channel frequencies, wide bit rate ranges, and preamble length possibilities.

The adoption of UWB technology will allow the application of distributed robotic systems [1]. In-house robotic applications such as home cleaning or warehouse transportation can offer last-mile delivery solutions. Smart manufacturing in Industry 4.0 must track the entire production process to always make the right decisions in real-time. Internal logistics at the factory floor level should lead to increased transparency, safety, and productivity. Positioning by UWB systems, when reaching precisions in the range of centimeters, uses triangulation or trilateration methods. Devices under UWB technology must be small to be able to develop portable equipment so as not to create fixed infrastructure which would reduce potential applications.

The advances achieved should bring new opportunities such as secure access control, device-to-device communications, as well as location-based services [1]. The automotive industry is implementing key management systems using mobile phones which allow access and start the operations of cars only when the digital key and the precise location of the phone match the user of the vehicle.

UWB systems are continuously introducing new integrated circuits to develop new algorithms and with-it new applications to exploit machine learning, collaborative positioning, and sensor fusion [1]. Sensor fusion seeks to maximize output information by computationally combining measurements from multiple sources. To make more reliable and precise estimates, it is necessary to merge several tracking systems into one UWB system. Machine learning makes complex tasks accessible by allowing computers to learn by themselves to perform tasks autonomously without needing to be programmed. Multilateration techniques use distant observables with direct signal propagation models to estimate the position of an agent. The estimation is more reliable by dynamic state models which use both current observations as well as previous positions using Bayesian statistics.

4. New developments in ultra-wideband (UWB) systems

The novel components that are being implemented by different research groups around the world for the optimization of ultra-wideband (UWB) systems are outlined in **Figure 1**. The use of integrated circuit design as well as materials science and engineering has a direct influence on the development of these new technological alternatives. These can be used for the implementation of ultra-wideband (UWB) systems.

An optical radar system based on a UWB resonator circuit using a microstrip line (MSL) has been proposed for monitoring the movement of human beings, whether healthy or sick [2]. The basic requirements of the design without degrading the quality factor (Q) for full bandwidth are a high signal-to-noise ratio, low power consumption, accuracy, and robustness. The resonator is implemented using an RLC circuit with all its components connected in parallel, having a photodetector at its input and an amplifier at its output, as shown in **Figure 2**. The resonator impedance is expressed according to Eq. 1, it must be kept almost constant at a value of 50 ohms with a bandwidth of 7.2 GHz when tuning the value of the inductor. The use of the microstrip allows a match of approximately 53.5 ohms which is close to the standard impedance in the range of 2–12 GHz. The inductor is implemented using a microstrip-based transmission line made of copper layers of a 65 nm process for radio frequency.

$$Z = R \,||\, Ls \,||\, \frac{1}{Cs} = \frac{RLs}{RLCs^2 + LCs + R} \tag{1}$$

The possibility of designing bandpass filters for UWB systems with cutoff bands at 5.18, 5.86, and 7.92 GHz has been explored [4]. Other filter properties that were achieved such as insertion loss of less than 1.5 dB, return loss of less than 15 dB, as well as high attenuation of unwanted in-band signals were also achieved. The filter structure is based on four stepwise controlled impedance open stubs using uniform transmission lines which provide the required tunability. The stubs have been inserted between two transmission zeros with the central stub in the middle of the uniform transmission line, where the stubs for each of the bands have been branched as illustrated in **Figure 3**.

A transmitter operating in the 3–5 GHz band for ultra-wideband (UWB) impulse radio applications using on-off keying based on a tunable memristor has been

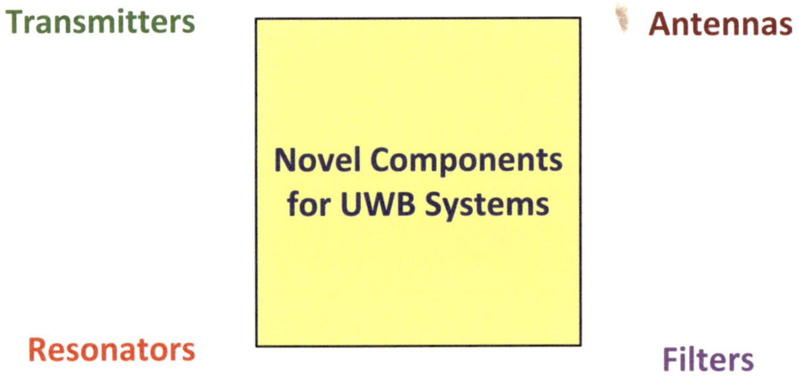

Figure 1.
Novel components for UWB technology.

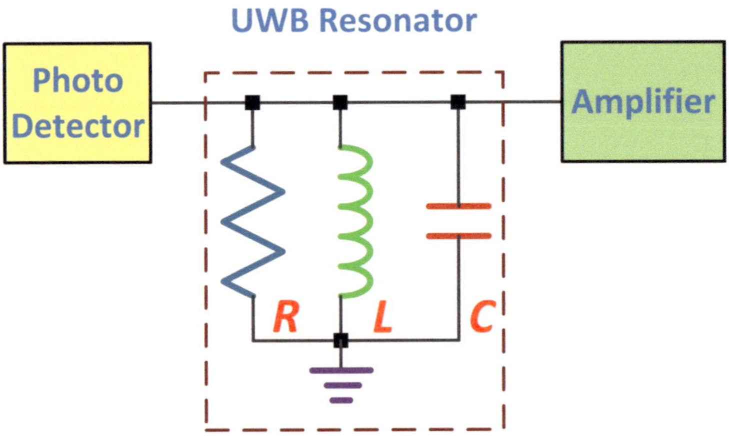

Figure 2.
Electrical model of a resonator for UWB systems.

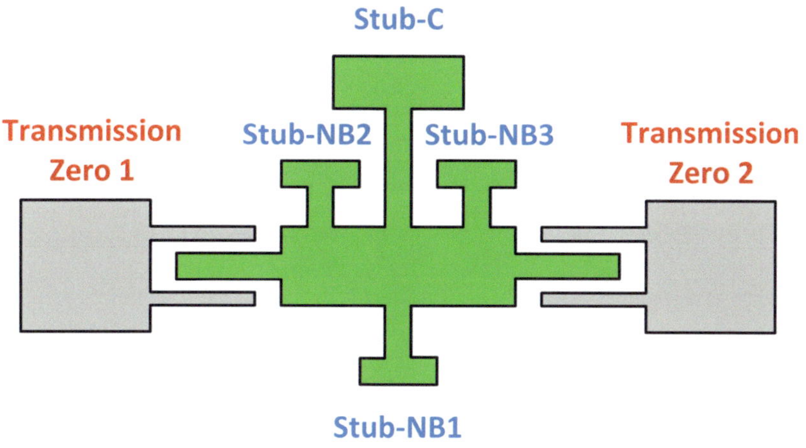

Figure 3.
The basic structure of a bandpass filter for UWB systems.

proposed [5]. The tuning opportunity comes from two external control signals that are used to modify the value of the memristance and the width of the output pulse to control the bandwidth and its center frequency, as depicted in **Figure 4**. The contribution of the memristor is associated with the reduction of power consumption, generating a wide bandwidth and a tunable power spectral density. This transmitter was implemented under standard 0.18-micron CMOS technology using a 1.8 V power supply. This approach is easy to implement and has flexibility that makes it attractive from its design. The output pulse width ranges from 0.8 to 1.74 ns using a control signal, the output excursion is 483 mV peak-to-peak, and it dissipates an energy of 9.48 pJ/pulse applying a pulse repetition frequency of 10 MHz through a 50-ohm load resistor.

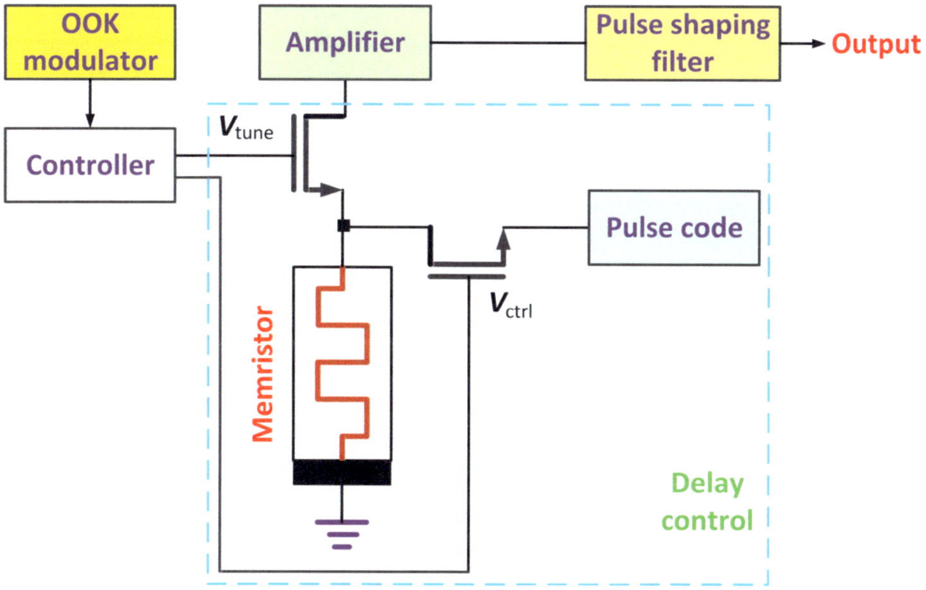

Figure 4.
Memristor-based transmitter for a UWB system.

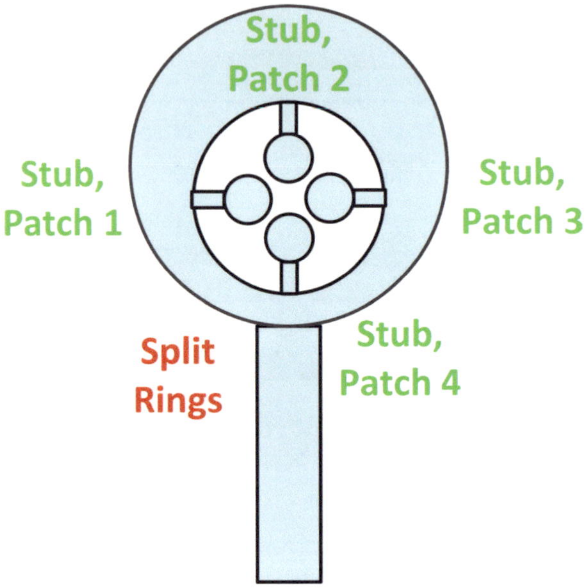

Figure 5.
Structure of a monopole antenna for a UWB system.

Another element of a UWB system that has been updated is the antenna. A planar monopole antenna operates in the range from 2.76 to 11 GHz in notch sub-bands in intervals from 3.75 to 4.81 and from 5.24 to 6.21 GHz, with a return loss of 10 dB [6]. These sub-bands are strategically placed to eliminate interference from other radio frequency bands. The tuning capability uses loaded stubs with four circular patches and two complementary split ring resonators operating as ground planes to achieve the cutoff bands. The implemented monopole antenna is shown in **Figure 5**.

A flared quasi-Yagi offset-fed monopole UWB antenna operating in the 3.06–12.37 GHz range has been implemented using an FR-4 dielectric material with dimensions in centimeters [7]. In addition, the antenna has superior bandwidth, reliability, speed, and high resolution, advantages that are usable for UWB technology. The reflection coefficient was less than −10 dB for the previously reported bandwidth. The maximum gain was 10.07 dBi for a frequency of 10.4 GHz with a peak radiation efficiency of 92.64% and with a radiation efficiency of 73%.

The use of microstrips for direct sequence ultra-wideband (DS-UWB) technology design has been raised [8]. This design uses planar structures with a wide bandwidth that exploits the impedance and pattern for applications in the C, X, and Ku bands using short pulses using radio pulse and exploited in multipath cases. This type of antenna is very interesting for systems with the Internet of Things (IoT), multiple inputs and multiple outputs (MIMO), as well as high-speed and short-distance communications.

5. Conclusions

The success of UWB systems over GPS systems is that in both indoor and outdoor environments, it is the smallest location error of the former. UWB systems perform most effectively under short pulses spread over a wide bandwidth. Narrowband signals from GPS systems degrade due to the multiple paths present. The accuracy of the GPS is compromised by the availability of repeaters and the time delay between repeaters and receivers.

Researchers worldwide will continue to look for design options that involve reduced complexity, a tendency to decrease power consumption, achieve sustainability, as well as tuning capabilities through reconfigurable options. The use of integrated circuit design as well as materials science and engineering has a direct influence on the development of novel technological alternatives. These can be used for the implementation of ultra-wideband (UWB) systems with better properties.

Acknowledgements

The author appreciates the support of the Instituto Tecnológico Superior de Irapuato to develop this research.

Thanks

The author wants to thank his wife and son for their support and time to edit this book. The author appreciates the support of Ana Cink working for IntechOpen as an author service manager.

Author details

Rafael Vargas-Bernal
Higher Technological Institute of Irapuato, Irapuato, Guanajuato, México

*Address all correspondence to: rvargasbernal@gmail.com

References

[1] Elsanhoury M, Mäkela P, Koljonen J, Välisuo P, Shamsuzzoha A, Mantere T, et al. Precision positioning for smart logistics using ultra-wideband technology-based indoor navigation: A review. IEEE Access. 2022;**10**:44413-44445. DOI: 10.1109/access.2022.3169267

[2] Gorre P, Vignesh R, Song H, Kumar S. A 64 dBΩ, 25 Gb/s GFET based transimpedance amplifier with UWB resonator for optical radar detection in medical applications. Microelectronics Journal. 2021;**111**:105026. DOI: 10.1016/j.mejo.2021.105026

[3] Waqar A, Ahmad I, Habibi D, Phung QV. Analysis of GPS and UWB positioning system for athlete tracking. Measurement: Sensors. 2021;**14**:100036. DOI: 10.1016/j.measen.2020.100036

[4] Taibi A, Trabelsi M, Saadi AA. Efficient design approach of triple notched UWB filter. AEU - International Journal of Electronics and Communications. 2021;**131**:153619. DOI: 10.1016/j.aeue.2021.153619

[5] Barraj I, Bahloul MA, Masmoudi M. Design of 3-5 GHz Tunable memristor-based OOK-UWB transmitter. AEU - International Journal of Electronics and Communications. 2021;**132**:153664. DOI: 10.1016/j.aeue.2021.153664

[6] Puri SC, Das S, Tiary MG. UWB monopole antenna with dual-band-notched characteristics. Microwave and Optical Technology Letters. 2020;**62**(3):1222-1229. DOI: 10.1002/mop.32112

[7] Nella A, Bhowmick A, Rajagopal M. A novel offset feed flared monopole quasi-Yagi high directional UWB antenna. International Journal of RF and Microwave Computer-Aided Engineering. 2021;**31**(6):e22653. DOI: 10.1002/mmce.22653

[8] Sarkar T, Ghosh A, Chakraborty S, Singh LLK, Chattopadhyay S. A new insightful exploration into a low profile ultra-wide-band (UWB) microstrip antenna for DS-UWB applications. Journal of Electromagnetic Waves and Applications. 2021;**35**(15):2001-2019. DOI: 10.1080/09205071.2021.1927855

Chapter 2

Probe-Fed Polygonal Patch UWB Antennas

Abhishek Joshi and Rahul Singhal

Abstract

The chapter deals with the design of probe-fed planar antennas to operate at wider bands and techniques to improve peak or boresight gain using reflectors. The phenomenon of frequency excitation in dual-band, that is, C-band and X-band using the technique of partial removal of the ground plane, is well demonstrated here. The impedance bandwidth achieved by the sample antenna is 285 MHz and 380 MHz, respectively. The reduced ground plane technique is further exploited along with modifications in the shape of the ground plane to cover the entire ultra-wideband (UWB) range in a probe-fed hexagonal monopole antenna. Due to the existence of higher modes and especially when fed with a probe, UWB antennas are only capable of providing mediocre gain at higher frequencies. An approach to increase the probe-fed hexagonal UWB antenna's peak gain involves the utilization of an appropriate reflector. The antenna is given an artificial magnetic conductor (AMC)-based reflector, which increases the peak gain as well as boresight gain across a band ≤ UWB. Peak and boresight gains of 3.74 dB and 5.5 dB, respectively, are observed with AMC. The equivalent circuit model and simulated impedance results of the sample antennas are validated with the measurement results.

Keywords: probe-fed, polygonal patch, UWB antennas, coaxially fed, AMC reflector

1. Introduction

Antennas are essential front-end entities of modern wireless communication systems. Today's wireless communication systems utilize planar antennas, which are popular over other types of radiators, especially in personal communication devices for wireless access. The reason for the popularity of planar antennas is essentially credited to their being economical, low profile, and easy integration with portable and personal communication devices. The advantages of planar antenna do accompany some drawbacks too, for example narrow bandwidth, poor radiation efficiency, and low power handling capacity. The limitations of patch antennas that need to be addressed in the present communication system are their intrinsic narrow-band behavior and polarization purity [1]. A signal transmitted by any radio base station cannot maintain its polarization state as it transverses toward the mobile terminal device due to channel properties. Thus, the polarization purity of antennas installed in terminal devices cannot be represented as a very strong design constraint [2]. It is difficult to anticipate the polarization state of the signal received on an antenna, even with the use of statistical analysis [3], suggesting that antennas should not be constructed based only on polarization assumptions for the reception. An antenna with a high cross-polarization level is a clear choice for polarization diversity applications and does not negatively affect radio-link performance as in cellular or satellite communications [4]. The necessity to keep the antenna dimensions as small as possible destroys the pretense of meeting greater bandwidth requirements. In the earlier generation of communication systems, the task is usually accomplished by techniques that rely on introducing slots or using reactive loads, but in current and future generation systems, subscriber expectation from service providers is extensive, and thus, novel solutions should be explored.

To guarantee that more services are available on various carrier frequencies, the possibility of a multi-frequency operation is expected in a communication link. Both wide-band and multi-frequency operations rely on the stimulation of two or more resonances, which in the case of wide-band antennas must resonate close together, while in the case of multi-frequency operation, they must resonate widely apart. Polygonal patch forms are an intriguing area for investigation among the patch antenna geometries that exhibit many resonances [5–9]. For an antenna designer to investigate and excite many resonances to obtain broadband or multi-resonant characteristics for the antenna, the patch size, the number of edges, the slope of the edges, etc. represent acceptable degrees of freedom [10, 11].

Complex electromagnetic wave interactions may be modeled using supercomputing in the frequency and time domains. It's challenging to create new antenna designs that can carry out demanding duties [12]. Kolundzija et al. [13] proposed an automated meshing of polygonal surfaces to segment a polygonal model into convex quadrilaterals to analyze efficiently as per electromagnetic theory. Sorokosz et al. [14] confirmed that a circular patch can be approximated with an appropriately designed polygon when the model analysis is performed. Many commercial simulation software have emerged during the past two decades, for example, computer simulation technologies (CST), Microwave Studio (MWS), Ansoft HFSS, etc., have supported the rapid advancement of polygonal patch antenna research. They have become fundamental tools in the design and simulation of planar antennas. The excitation and boundary conditions are satisfied via the method of moments (MoM), which makes use of integral equations that are discovered for the fields generated by unidentified currents. Maxwell's equations are transformed into difference equations via the full wave (FW) approach or the finite-difference time-domain (FDTD) method. The Rayleigh–Ritz variational approach is used by the finite element method (FEM) to solve Maxwell's equation as the vector wave equation [15]. Yikai et al. [16] reviewed characteristic modes for radiation problems from antenna design to feeding design and found that with the help of many unique and attractive features of control management, physical understandings of the radiating problems can be much clearer, computation burdens in antenna optimization procedure can be greatly alleviated, and designs with favorite features such as compact and low profile can often be obtained. After a real prototype has been built and measured in a lab, the CST MWS Suite enables virtual prototyping of one's idea, reducing any unpleasant surprises.

Various techniques are available for the fabrication of microstrip patch antenna, but a commonly popular economical fabrication technique such as the wet-etching method will be more suitable in case one of the objectives is technology transfer to the industry shortly. Photolithography technique can be used to transfer the mask image on electroplated copper on a printed circuit board (PCB) using negative photoresist, and then, it is developed in a developer solution. The unexposed unwanted features are etched out in an etching chemical solution.

A polygon with an odd number of edges as in the case of pentagon geometries may be of a symmetrical shape as that of a regular hexagon with an even number of edges or can be asymmetric too. After the substrate thickness and permittivity are chosen, the resonant frequencies of a regular patch rely on the geometry of the conductor. By using the appropriate degrees of freedom, the resonances along the frequency axis may be controlled [11]. Polygonal patch geometries can be triangular or rectangular as per the conventional definition of the word *polygon,* but in general conception, polygon refers to geometries with edges greater than or equal to five [17]. The text in the chapter will follow this general conception.

The selection of a dielectric substrate for a patch antenna is one of the significant aspects of the antenna design. Many researchers have exploited different dielectric substrates with different permittivities for patch antenna in different applications. Radiofrequency (RF) energy may be provided to microstrip patch antennas using several methods. Contact feed and non-contact feed are two categories into which these approaches may be divided. Microstrip line and coaxial probe feeding are the most often used contact feeding methods, while aperture coupling and proximity coupling feeding are the most widely used non-contact methods. Distinct geometries of patch antenna like a triangular, rectangular, pentagon, hexagon, etc. are well explored by many researchers. By altering the substrate's dielectric constant, the patch's sizer and the conductor strip's metal thickness, any antenna design that is acceptable for operation across a range of impedance bandwidths may be produced. Hexagonal is a popular geometry used to design a stripline-fed monopole antenna [18–21] and is investigated in detail by Ray et al. [22]. There are dozens of research publications about microstrip patch antennas scattered among traditional periodicals. In a 2012 assessment of patch antenna development strategies, Lee et al. [15] discovered that all the ways that described widening the working bands of patch antennas result in increased volume, negating the low-profile benefit of microstrip patch antennas. Due to its broader band yet higher order modes, the hexagonal structure is favored over other geometries [22, 23]. Further use of the hexagonal shape may be used to produce lower modes and produce broader bands or UWBs.

Ground plane geometry plays a vital role in the design of polygonal patch antenna. For contemporary wireless communication systems, hexagonal monopole antennas fed at the edge and the vertex are extensively investigated at lower frequencies [24]. Because CPW-fed antennas cannot be used for the small overall construction, efficient direct probe feeding becomes the best option. However, employing a coaxial probe or connection to feed a direct-fed UWB monopole antenna presents a problem for antenna researchers. Antenna feed plays a significant role in exciting higher order mode and modification of the feed; such as using a larger diameter probe may permit the antenna to generate a higher order mode [25]. Even with a modified connector feed, the design of a probe-fed UWB planar antenna is still a challenge for the antenna research community. According to [26], adding more ground plane structures may raise the boresight gain of planar antennas; however, the strategy also results in larger antennas. For a stable and reliable radiation pattern, a technique for producing a directed pattern in a hexagonal UWB antenna fed via probe must be understood. The UWB monopole antenna's peak gain may be increased using a variety of methods, including a reflector with a frequency-selective surface (FSS) base. The gain of the patch antenna at higher frequencies within the operational range of the antenna may be greatly increased by defects in the ground plane, such as the insertion of alternative slot geometries or alteration of the ground plane shape.

The literature survey concludes that although many researchers have explored polygonal geometries through their work, a systematic approach to understanding the effect of polygonal geometry over antenna performance has not been undertaken in the past and still is an interesting subject of research. The main limits of polygonal patch antennas in a present communication system are both the intrinsic narrow-band behavior and the polarization purity. The request to keep the antenna's overall dimensions small is often made in conjunction to meet these increased bandwidth requirements.

Every novel patch antenna that should be designed, developed, and characterized should achieve some of the common patch antenna features such as compact antenna size, operating band enhancement, gain enhancements, etc. To reduce the size of the antenna by moving the resonance frequency to the lower side, the ground plane defects and fractal geometries should be investigated. A printed circuit board with FR-4 as a dielectric will be an economic approach for designing the polygonal patch antenna. Recent studies point toward the use of coaxial feed as one of the approaches that may yield multiband or broadband antenna characteristics, which may be further exploited to achieve super wide-band characteristics. The literature review reports various attempts to address polygonal patch antenna(s) and to explore it more.

Based on the available literature, the motivation of the chapter is briefed as follows:

- Analyzing a hexagonal geometry, which reveals that antenna performance needs further analysis and still is an interesting subject of research.

- Exploring limits of hexagonal patch antennas such as the intrinsic narrow-band behavior and the polarization purity.

- Keeping the antenna's overall dimensions small in conjunction with increased bandwidth demands.

- Utilizing an economic printed circuit board with FR-4 as a dielectric for designing the hexagonal patch antenna.

Using a survey of reported literature and preliminary work including hexagram design [27], pentaflake antenna [27], and polygonal patch antenna, it is observed that a probe-fed polygonal patch antenna is not much explored, with coaxial feed as one of the approaches that may yield multiband or broadband antenna characteristics, which may be further exploited to achieve super wide-band characteristics.

2. Low cross-polarization vertex-fed hexagonal antenna

Impedance mismatch affects probe-fed hexagonal patch antennas, particularly when the feed is situated near one of the polygon vertices. High cross-polar levels also affect hexagonal planar antennas. A technique for the low cross-polarization vertex-fed hexagonal antenna is demonstrated. To indicate improvement in impedance values, vertex feeding is illustrated in this section. It has an impedance that is excessive when compared to the probe impedance. The suggested approach may be tuned to match the impedance and obtain excellent return loss. Here, a wide bandwidth (600 MHz) low cross-polarization vertex-fed hexagonal antenna operating at a frequency of 5 GHz is suggested [28]. Therefore, the antenna shown in **Figure 1** is a good fit for unlicensed UNII-1 indoor wireless local area network (LAN) applications. Antennas 1–3 (A1, A2, and A3) are similar in structure as shown in **Figure 1** but differ in their geometrical features, which are listed in **Table 1**. These developed antennas will demonstrate the impact of ground plane miniaturization.

The co- and cross-polarization levels are significantly influenced by the substrate dimension [29]. All three antenna designs' substrate dimensions as presented in **Table 1** are set at a value that is optimal to reduce cross-polarization. The antennas

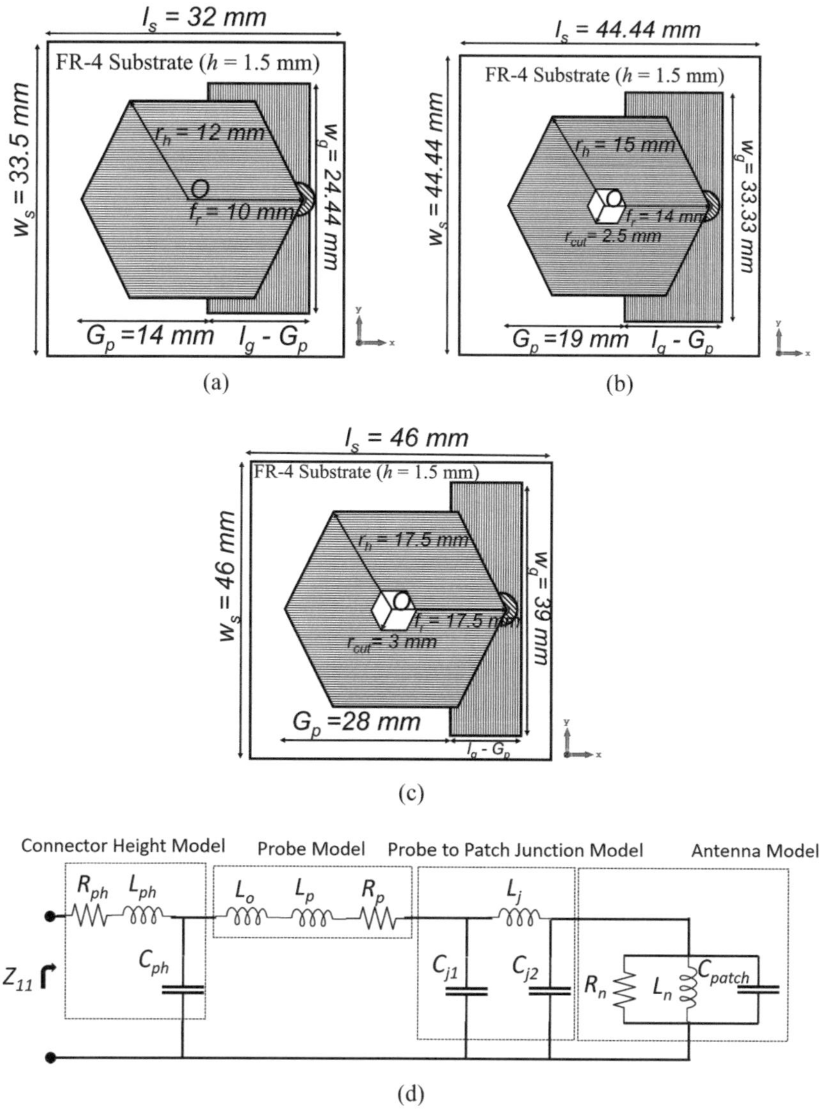

Figure 1.
Vertex-fed hexagonal antenna with the reduced ground for (a) A1, (b) A2, (c) and A3. (d) Circuit model of the antenna with its probe feeding network.

Design considerations (mm)	Antenna		
	A1	A2	A3
Ground width (w_g)	24.44	33.33	39.00
Hexagonal patch circumradius (r_h)	12.00	15.00	17.50
Ground reduction factor (G_p)	14.00	19.00	28.00
Effective ground length (l_g - G_p)	10.44	14.24	9.00
Feed point radius from O (f_r)	10.00	14.00	17.50
Slot radius (r_{cut})	0.00	2.50	3.00
Substrate length (l_s)	32.00	44.44	46.00
Substrate width (w_s)	33.50	44.44	46.00

Table 1.
Parameters used for three antenna designs.

may be modeled using a simple *RLC* resonant equivalent circuit [17, 30] as shown in **Figure 1d**. According to formulae, lumped elements (*RLC*) may be derived for all the antennas through Eqs. (1)–(3). Eqs. (1)–(3) can be used to calculate patch capacitance (C_{patch}), patch inductance (L_n), and patch resistance (R_n) for a given f_n, and they can also be used to calculate Z_{11} and patch impedance (Z_{patch}), which are expressed in Eqs. (4) and (5), respectively.

$$C_{patch} = \left(\frac{\varepsilon_0 \varepsilon_r}{2h}\right)\left(\frac{3\sqrt{3}r_h^2}{2}\right)\left(1 - \frac{G_p}{l_g}\right) \tag{1}$$

$$L_n = \frac{1}{(2\pi f_n)^2 C_{patch}} \tag{2}$$

$$R_n = \frac{Q}{(2\pi f_n) C_{patch}} \tag{3}$$

where f_n is the working band's center frequency, and h and r_h are the properties of the substrate, respectively. G_p affects the overlapping area of the patch with the ground, resulting in the change of patch capacitance (C_{patch}). The input impedance (Z_{11}) as shown in **Figure 1d** with the probe feeding network is given by the following expression.

$$Z_{11} = R_{ph} + j2\pi f L_{ph} + \cfrac{1}{\cfrac{1}{R_p + \cfrac{1}{\cfrac{1}{\cfrac{1}{\cfrac{1}{Z_{patch}} + j2\pi f C_{j2}} + j2\pi f L_{j1}} + j2\pi f C_{j1}} + j2\pi f L_0 + j2\pi f L_p} + j2\pi f C_{ph}} \tag{4}$$

An equivalent circuit can be modeled along the same lines as in [30] and is presented in **Figure 1d**. Due to the probe-to-patch junction, the capacitance (C_{j1} and C_{j2}) and inductance (L_{j1}) are present. When the probe is within the substrate, it displays resistance (R_p) and inductance (L_o + L_p). Resistance (R_{ph}), inductance (L_{ph}), and capacitance (C_{ph}) are introduced depending on how high the probe is above the substrate.

$$Z_{patch} = \frac{1}{\frac{1}{R_n} + j2\pi f C_{patch} - \frac{j}{j2\pi f L_n}} \tag{5}$$

then, Eq. (6) may be used to compute the reflection coefficient, S_{11} (dB).

$$S_{11}(dB) = 20\,log_{10}\left(\frac{Z_{11} - Z_0}{Z_{11} + Z_0}\right) \tag{6}$$

where Z_0 is the characteristics impedance of the probe, which is 50 Ω. Eq. (6) may be used to generate the $|S_{11}|$ (in dB) of the resonant equivalent circuit model shown in **Figure 1d** by simply sweeping the frequency (f) for the desired frequency range. The reflection coefficient generated using the resonant equivalent circuit provides a similar result as obtained using simulation software as shown in **Figure 2** for (a) A1, (b) A2, and (c) A3.

The value of G_p is varied to analyze its effect on the reflection coefficient of the three-antenna designs used for the analysis and presented in **Figure 2a–c**, respectively. The following can be concluded by observing the results presented in **Figures 2** and **3**, that is, simulated and measured, respectively. A ground reduction method is appropriate for matching the impedance of the demonstrated antenna, thereby suppressing any extra resonance. The suggested antenna impedance (53.37-j-5.2) matched at 5 GHz. The Antenna 3 is appropriate for WLAN (UNII-1) applications since it has a 3 dB gain and works at 5 GHz within a 600 MHz bandwidth. Further details regarding low cross-polarization for the vertex-fed hexagonal antenna at the antenna's boresight may be found in [28], which suggests that antenna 3 ground plane reduction suppresses higher order mode.

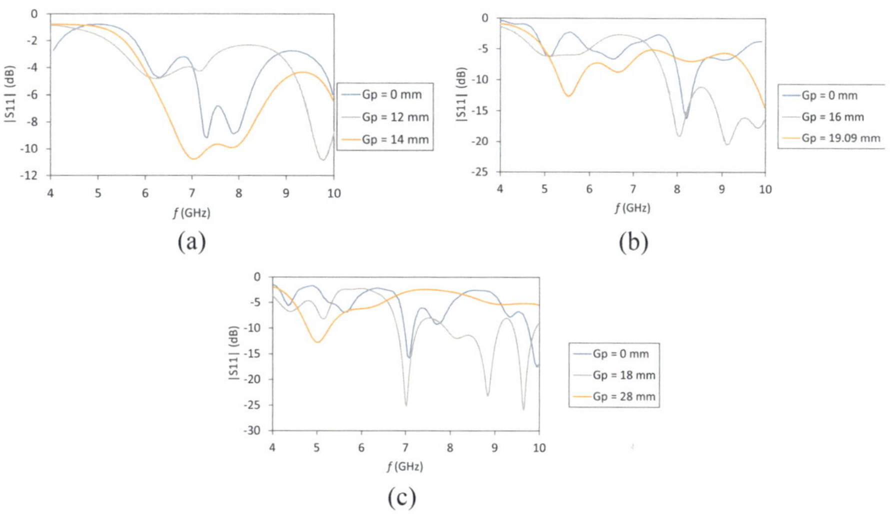

Figure 2.
Simulated reflection coefficient, S_{11} (dB), for (a) A1, (b) A2, and (c) A3.

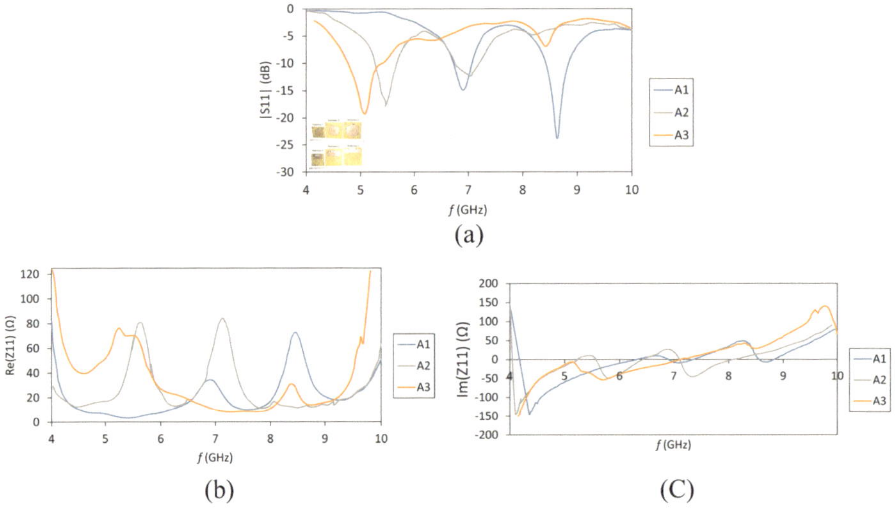

Figure 3.
(a) $|S_{11}|$ (in dB), measured return loss (Inset: front and back of the developed antennas). Variation in the antennas' input impedance: (b) real part, (c) imaginary part.

3. Reduced ground plane probe-fed polygonal patch for *C*- and *X*-band applications

This section analyzes and presents the effects of a decreased ground plane on radiation and impedance in a hexagonal antenna that is supplied coaxially. In this study, the measured and simulated impedance findings for an antenna design are compared with the equivalent circuit model. Lower *X*-band and higher *C*-band frequencies that are stimulated by the ground plane are reduced. Here, the antenna displayed in **Figure 1a** and described in an earlier section of the chapter is again used for the analysis. The developed hexagonal antenna makes use of a substrate that measures 32 × 33.5 mm^2 and a smaller ground plane that is 10.44 ($l_g - G_p$) × 24.44 (w_g) mm^2 as shown in **Figure 4** [31]. **Table 1** for antenna A1 may be referred to for the description and values of all the design variables presented in **Figure 4(a)**. The suggested antenna has an impedance bandwidth in the C-band and X-band of 285 MHz and 380 MHz, respectively. For *C*-band and *X*-band applications, the developed antenna with a smaller ground plane performs well.

As illustrated in **Figure 5**, *RLC* components are employed to mimic the proposed antenna and the corresponding circuit. As mentioned in the third part of the study, the equivalent circuit model is produced for the resonating frequencies of 7 GHz (f_1) and 8.69 GHz (f_2) as observed during experiments on a vector network analyzer (VNA). The values of the lumped component *RLC* are derived like that used in [32] for an E-patch. For a hexagonal patch antenna, Eqs. (7)–(9) are found by changing the equation provided in [33].

$$C' = C_n = \frac{\varepsilon_o \varepsilon_\varepsilon A_e}{2\mathrm{h}} \tag{7}$$

$$L_n = \frac{1}{(2\pi f_n)C'} \tag{8}$$

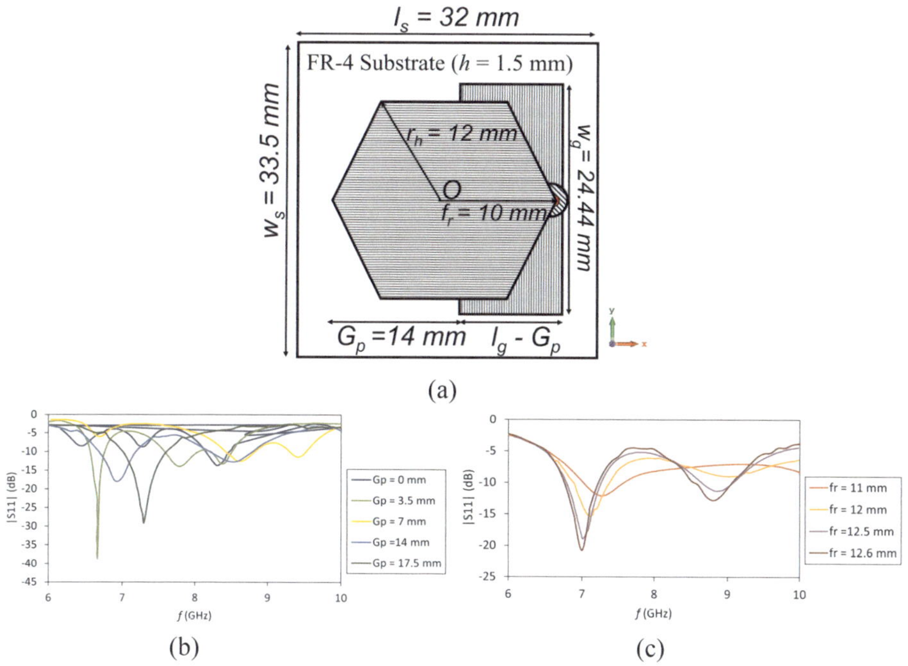

Figure 4.
(a) Reduced ground plane probe-fed polygonal patch for C and X-band applications. (b) Reflection coefficients (S_{11}) when G_p and f_r = 12.6 mm. (c) Simulated S_{11} (dB) of the antenna when f_r varies when G_p = 14 mm.

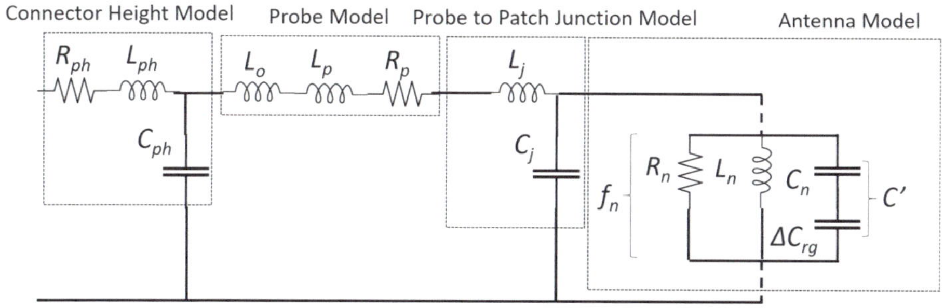

Figure 5.
Circuit model of the antenna with its probe feeding network.

$$R_n = \frac{Q}{(2\pi f_n)C'} \tag{9}$$

where A_e is a patch area over the ground plane, C' and C_n are calculated using Eq. (7), and n = 1, 2, and so on.

For a complete ground and capacitance of C_n = 4.75 pF (for frequencies f_1 and f_2) as shown in **Figure 5**, the patch area, A_e, is 374.12 mm^2. The patch size, A_e = 150.03 mm^2, and capacitance, C' = 1.90 pF, are for decreased ground (G_p = 14 mm). The value of the capacitance for the hexagonal patch, which is represented by its series capacitance (ΔC_{rg}), is decreased as a result of the reduction in the ground. The formula in Eq. (10) may be used to get a hexagonal patch's extra series capacitance.

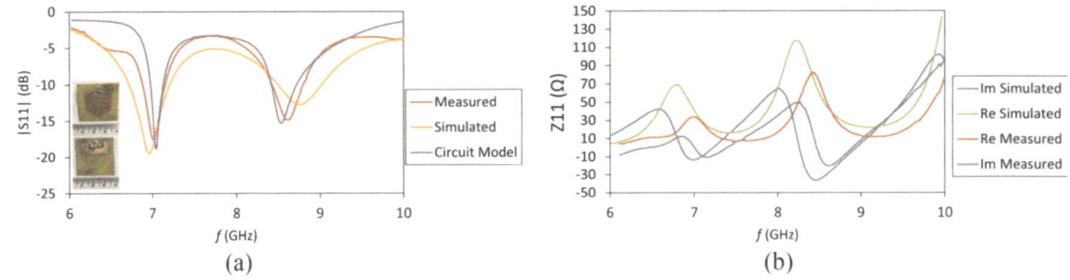

Figure 6.
(a) S_{11} (in dB), the reflection coefficient of the antenna. (b) Measured real and imaginary impedance (Z_{11}).

$$\frac{1}{C'} = \frac{1}{C_n} + \frac{1}{\Delta C_{rg}} \tag{10}$$

VNA is used to measure the antenna's reflection coefficient S_{11} (dB), which is displayed in **Figure 6**. In **Figure 6**, simulation results utilizing CST MWS and feed probe data from the *RLC* model are compared. By observing the measured, simulated, and circuit model bandwidth, it may be concluded that, due to the inductance introduced by the coaxial feed, the VNA bandwidth is lower (SMA connector).

To analyze the impedance matching, the real and imaginary impedance parts are measured and compared with simulated results as shown in **Figure 6**. By using values simulated of real and imaginary impedance, it is shown that the optimal value of input impedance, Z_{11}, at 8.6 GHz is 49.16 - j 27.19 in the operational range for the hexagonal form (**Figure 6b**). In the operational range of the designed antenna, measurements observed on a VNA reveal that the value of Z_{11} measured at 8.62 GHz is 50.06 - j 18, which is pretty much like the value described above or observed during simulations.

To attain an *X*-band lower frequency of 8.69 GHz, the impact of the ground reduction in the hexagonal design is investigated and assessed. By raising G_p from 0 to 18 mm, the ground is reduced. Due to direct probe feeding, inductance is added, reducing the antenna impedance bandwidth. At 7 GHz and 8.69 GHz, the input impedance, Z_{11}, is measured to be 38.6 + j 2 and 50.06 - j 18, respectively. In the operational bandwidth, the radiation patterns are frequency independent. The antenna shown in this section is suitable and affordable for wireless applications since it has a bandwidth of 285 MHz and 340 MHz at frequencies of the *C*-band and *X*-band, respectively. The study in this section is limited to impedance analysis. Further analysis such as radiation pattern and gain of the presented antenna is given in [31].

4. Flangeless SMA connector hexagonal UWB antenna fed via probe

This section of the chapter introduces a UWB antenna that is supplied via a flangeless standard SMA connection close to one of the hexagonal patch's vertices as reflected in **Figure 7**. **Figure 7a** uses the same notation as described in **Table 1** to describe the design variables. To accomplish UWB with monopole radiation characteristics, the antenna has a ground plane that is half elliptical but truncated and has a rectangular slot. The antenna prototype has a WLAN band rejection of 1.6 GHz from 4.9 GHz to 6.5 GHz and an impedance bandwidth of 8.3 GHz from 2.3 GHz to 10.6 GHz [34]. The removal of flanges transforms a *C*-band antenna into a UWB

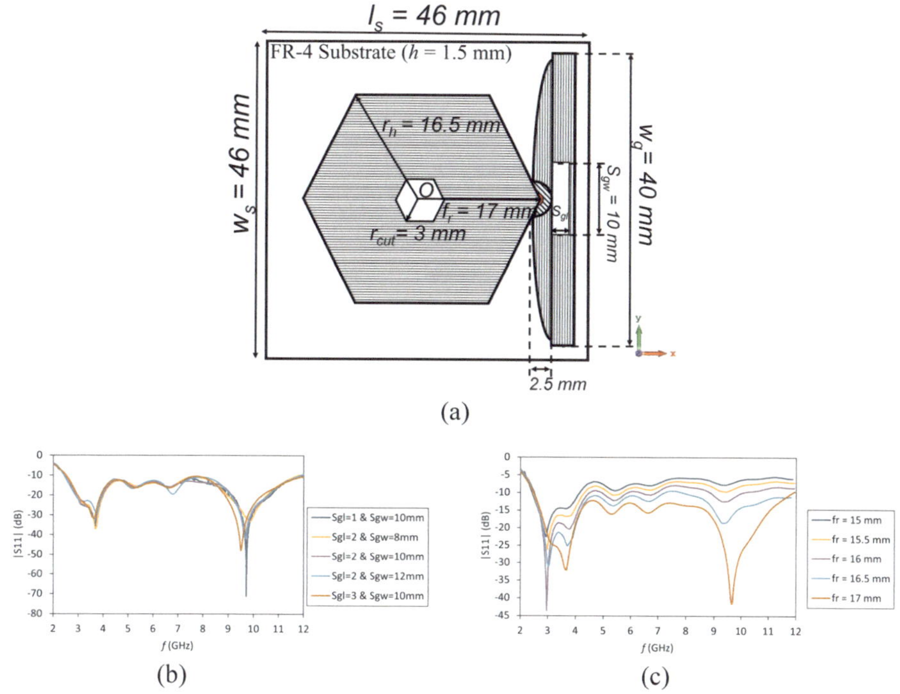

Figure 7.
UWB hexagonal antenna: (a) Dimensions and layout (c) Simulated $|S_{11}|$ *(in dB), when the feed point* (f_r) *is at 17 mm, and the hexagonal slot radius* (r_{cut}) *is 3 mm, for various values of slot width* (s_{gw}) *and slot length* (s_{gl})*. (d) Simulated scattering parameter,* $|S_{11}|$ *(in dB) vs. frequency, for various* f_r *value slot points* (f_r) *is at 17 mm, and* r_{cut} *is 3 mm when slot length* (s_{gl}) *is 2 mm and slot width* (s_{gw}) *is 10 mm.*

antenna, according to antenna tests. The proposed method may be used with a probe-fed antenna to obtain UWB radiation. Measurement results are consistent with what is anticipated based on simulation outcomes.

It is shown how an antenna presented in this section responds to SMA connection flanges in terms of impedance bandwidth. The suggested antenna displays *C*-band characteristics when fed via a connector with a flange, but a flangeless connector is a good option to obtain UWB characteristics in a direct-fed antenna. The designed antenna prototype shown in **Figure 8** exhibits WLAN band rejection between 4.9 GHz and 6.5 GHz. The designed antenna prototype has a 1.6 GHz WLAN spectrum rejection and is suited for UWB applications between 2.3 GHz and 10.6 GHz.

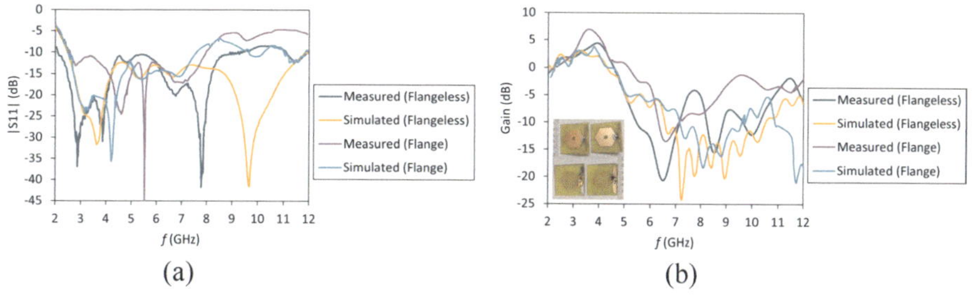

Figure 8.
Probe-fed hexagonal monopole UWB antenna. (a) Measured $|S_{11}|$ *(in dB) and (b) gain (dB).*

5. Determination of band edge frequencies of probe-fed printed hexagonal monopole antenna

The calculation of the lowest edge resonance frequency of a printed monopole antenna is not addressed much as done for the resonant frequency of a dipole antenna. It is well known that expressions available for the calculation of the resonant frequency of a dipole antenna cannot be used for the calculation of a lower edge frequency of a printed monopole antenna. Resonant modes in a dipole antenna are generated due to half-wave variation along its horizontal and vertical axis, but printed monopole antennas have negligible patch capacitance. In this chapter, an empirical formula is proposed to calculate the lower and higher edge frequencies of a probe-fed printed monopole antenna as it possesses band-pass impedance characteristics. Three types of printed monopole antennas have been studied and simulated for validation of the empirical formula proposed utilizing full-wave simulation software. It is observed that the probe-fed hexagonal antenna exhibits a wider band, which motivated us to validate the empirical formula through an experiment. Experimental results validated the results obtained from the proposed empirical formula. Percentage error magnitude is also calculated and presented in the section for each case studied in this work.

The empirical formulas for the calculation of the lower edge frequency of the impedance bandwidth (f_L) for a rectangular monopole antenna, a hexagonal monopole antenna, and a circular monopole antenna are given in [22, 35, 36], respectively. A stripline-fed quarter-wave hexagonal monopole antenna is modeled in [22], and the empirical formula to calculate the f_L (in GHz) of a stripline-fed hexagonal monopole antenna is given in [22]. Reference [36] exploits a modified trapezoid-based empirical formula given in [37] to derive the formula for a circular monopole antenna. The expressions given in [22, 35–37] can be used to calculate the f_L (GHz) of the designed antenna with a stripline feed when fed at the vertex of the printed monopole antenna. As the area of a hexagon is treated as an equivalent circular area, the empirical value of 1.15 is multiplied in the denominator to calculate the f_L of the vertex-fed hexagonal monopole. Also, instead of a quarter wavelength, that is, 0.25 λ, 0.24 λ is used in the expression, which results in a constant of 7.2 in the numerator [22]. Antenna configurations with different patch geometries, that is, rectangle, hexagonal, and circular monopole antenna with dimensions as given in [22, 35, 36], are used to calculate f_L and indicated in **Table 2**.

The expressions given in [22, 35–37] are modified by changing the parameters to avoid loss of generality, and f_L can be calculated near practical results. In the case of a printed monopole antenna, the lower edge frequency is a significant parameter rather than the resonant frequency. The lower edge frequency of a printed monopole antenna is given by

Antenna configuration	Measured/simulated f_L (GHz)	Calculated f_L (GHz) using			Percentage error in f_L using		
		[35]	[22]	[36, 37]	[35]	[22]	[36, 37]
Rectangle in [35]	1.5	1.35	0.92	1.55	10	38.8	3.3
Hexagon in [22]	1.1	1.06	1.0743	0.8975	3.8	2.336	18.41
Circular in [36]	2.65	3.40	2.54	2.65	28.3	4.2	1.766

Table 2.
Comparison of f_L *of various monopole antenna configurations.*

$$f_L = \frac{c}{\lambda\sqrt{\varepsilon_{eff}}} \tag{11}$$

where c is the speed of light.

But here, quarter-wave monopole antennas are used and compared. Thus, the length will be $L = \lambda/4$, which results in $\lambda = 4L_{eff}$; therefore,

$$f_L = \frac{c}{4L_{eff}\sqrt{\varepsilon_{eff}}} \tag{12}$$

$$f_L = \frac{7.5}{L_{eff}\sqrt{\varepsilon_{eff}}} \tag{13}$$

In the case of a stripline-fed monopole antenna, the effective length will be $L_{eff} = L + L_s$, and hence, the lower edge frequency is given by

$$f_L = \frac{7.5}{(L + L_s)\sqrt{\varepsilon_{eff}}} \tag{14}$$

where effective dielectric constant, $\varepsilon_{eff} = (\varepsilon_r - 1)/2$, ε_r is the permittivity of the substrate material, and L_s is the length of the stripline.

The empirical formula for the calculation of the f_L for circle and hexagonal monopole antenna is modified by considering $L = 2 \times R$ *and* $L = 2 \times h_r$, respectively, where R (in cm) is the circle radius and h_r (in cm) is the circumradius of the hexagon; the final form of the expression is as given in Eq. (15).

$$f_L(\text{GHz}) = \begin{cases} \dfrac{7.5}{(L + L_s)\sqrt{\varepsilon_{eff}}}; \text{for rectangle} \\ \dfrac{7.5}{(2h_r + L_s)\sqrt{\varepsilon_{eff}}}; \text{for hexagon} \\ \dfrac{7.5}{(2R + L_s)\sqrt{\varepsilon_{eff}}}; \text{for circle} \end{cases} \tag{15}$$

The empirical formula for f_L calculation of the circular monopole antenna as given in [36, 37] is modified for the hexagonal monopole antenna by considering circumradius $W_1 = W_2 = 3 \times h_r$ and $L = 2 \times h_r$, and the final expression for rectangle, hexagon, and circle are given by Eq. (16). Measured and calculated values are displayed in **Table 3**.

Antenna configuration	Measured/simulated f_L (GHz)	Calculated f_L (GHz) using the expression		Percentage error in f_L for the expression	
		(15)	(16)	(15)	(16)
Rectangle in [35]	1.5	1.46	1.55	2.7	3.3
Hexagon in [22]	1.1	1.107	1.11	0.6	0.9
Circular in [36]	2.65	3.51	2.65	32.4	1.766

Table 3.
Comparison of f_L of various monopole antenna configurations using modified formula.

$$f_L(\text{GHz}) = \begin{cases} \dfrac{904}{4\pi L + W_1 + W_2}; \text{for rectangle} \\ \dfrac{904}{8\pi h_r + 6h_r}; \text{for hexagon} \\ \dfrac{904}{10\pi R + 120h}; \text{for circle} \end{cases} \tag{16}$$

For the formulation of probe-fed hexagonal monopole antenna, the empirical value, that is, k (dielectric constant), is avoided by assuming it as '1', which is earlier used in stripline-fed hexagonal monopole antennas, due to the fringing extension and the effective dielectric constant ($1 < \varepsilon_{eff} < \varepsilon_r$ [38]). The additional effective h_{reff} and feed line length are also avoided since a direct probe is used to feed the vertex of the hexagon. Another reason for choosing $k = 1$ is that due to the negligible capacitance of the patch because of a monopole configuration and purely inductive patch, the effective dielectric constant leads to 1. The hexagonal monopole configuration is conventionally modeled as an equivalent circular monopole antenna. For a vertex-fed hexagonal monopole antenna, the length of the hexagon will be twice the circumradius of the hexagon, that is, $L = 2 \times h_r$; substituting it in Eq. (12) results in Eq. (17).

$$f_L = \frac{c}{4(2h_r)\sqrt{1}} \tag{17}$$

Although expressions for vertex-fed hexagonal monopole have been explored earlier and empirically [22], derived for a stripline-fed antenna, here an empirical formula that is more suited to a probe-fed hexagonal monopole antenna when fed at the vertex of the hexagon is presented:

$$f_L = \frac{7.5}{2h_r} \text{ or } \frac{3.75}{h_r} \tag{18}$$

where h_r is in cm. The empirical expression can be justified by the fact that the monopole can be modeled as a pure inductance with negligible capacitive effect.

Moreover, the higher edge frequency of the impedance bandwidth (f_H) of the probe-fed hexagonal antenna can also be estimated using the following empirical formula, that is, Eq. (18). A similar technique is used in [21] for the calculation of the f_H of an irregular hexagon. The higher edge frequency empirical formula is because the smallest edge of the patch will contribute to the wavelength; that is, the lowest radiating wavelength will be equal to the edge of the patch with the smallest dimension. But, in a regular hexagon, the edge dimension is the same as the circumradius of the hexagon. The estimation of f_H using Eq. (18) for the probe-fed hexagonal monopole antenna overlooks weakly rejected bands. Although the monopole antenna exhibits high pass impedance characteristics [35], the calculation of f_H is sometimes appreciated while designing UWB monopole antennas.

$$f_H = \frac{c}{h_r\sqrt{\varepsilon_{eff}}} \tag{19}$$

where c is the light speed in free space, h_r is in meters, and $\varepsilon_{eff} = (\varepsilon_r + 1)/2$.

The effective dielectric constant in Eqs. (18) and (19) is estimated from enormous simulations, experiments, and analyses to achieve an appropriate empirical formula for a probe-fed hexagonal monopole antenna. Eq. (18) is used to calculate the f_L for three planar monopole antennas, that is, square, circle, and hexagon. All three antenna configurations are assumed to have the same circumradius of 16.5 mm. The calculated values of f_L are indicated in **Table 4**. The values of f_L are further verified through CST MWS simulation results.

Three different probe-fed monopole antenna configurations, that is, rectangle, circle, and hexagon with feed position at the vertex of the polygon, are designed in CST microwave studio as shown in the inset of **Figure 9a** and simulated for S_{11} characteristics of the designed antenna as reflected in **Figure 9a**. The probe at the vertex of the polygon helps in designing the monopole antenna by avoiding overlapping with the ground. The patch has been modeled as a regular design with the circumradius h_r. The three designed configurations consist of an FR-4 substrate of 46×46 mm^2, a ground plane with dimensions 3×40 mm^2, and a patch with a circumradius of 16.5 mm. The ground dimensions of the antenna configurations are chosen such that the designed probe-fed hexagonal monopole antenna has a minimum size. All three antennas have almost the same lowest edge frequency as may be observed from **Figure 9** at -10 dB and as indicated in **Table 4**.

Eq. (18) is found to be more suitable for a probe-fed hexagonal monopole antenna, especially when fed at the vertex of the hexagon, which yields f_L = 2.27 GHz, which provides an error of 1.3% as indicated in **Table 4**. As observed from **Figure 10**, the probe-fed hexagonal monopole antenna possesses wideband characteristics and was further chosen for fabrication to design a probe-fed UWB monopole antenna. The hexagon monopole antenna shows weak rejection at 4.3 GHz as depicted in **Figure 10**.

Antenna configuration	Simulated		Calculated		Percentage error in	
	f_L (GHz)	f_H (GHz)	f_L (GHz) using expression (18)	f_H (GHz) using expression (19)	f_L	f_H
Rectangle	2.3	11.2	2.27	11.2	1.3	0
Circle	2.4	8.9	2.27	11.2	5.41	25.8
Hexagon	2.3	11.6	2.27	11.2	1.3	3.4

Table 4.
Comparison of f_L *of proposed probe-fed monopole antenna configurations using proposed formula.*

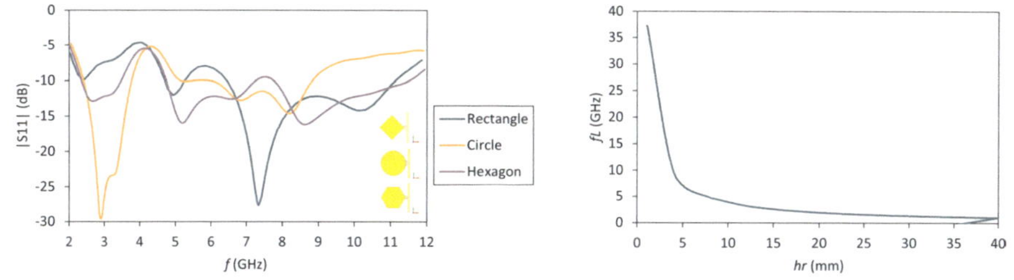

(a) (b)

Figure 9.
(a) Scattering parameter, $|S_{11}|$ (in dB) of different monopole antenna configurations (Inset: pictures of antennas). (b) Variation in f_L *with* h_r *of probe-fed hexagonal monopole antennas.*

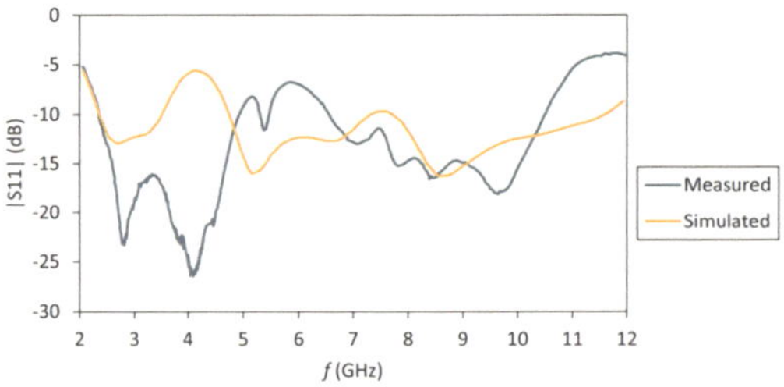

Figure 10.
$|S_{11}|$ (in dB) for probe-fed hexagonal monopole antennas.

The value of f_L is calculated for different values of h_r using Eq. (18) and observed using CST software for probe-fed hexagonal monopole antennas. The optimized value of h_r of probe-fed hexagonal monopole antennas demonstrates maximum bandwidth, for a given f_L, and is displayed in **Figure 9b**. The probe-fed hexagonal monopole antenna, especially when fed at the vertex of the hexagon, demonstrates maximum bandwidth because of the best transition of impedance bandwidth.

f_L is found to be 1.72 GHz when h_r = 1.65 cm using Eq. (14). The dimensions of the hexagon mentioned earlier are chosen to accommodate the entire S-band along with the UWB band. But, after fabrication, the f_L calculated using Eq. (14) provides an error of 25.21%, because probe feeding is used for the excitation of the hexagonal monopole antenna. A more suitable expression for the calculation of f_L for probe-fed hexagonal monopole antenna is presented in Eq. (17), and it provides an error of only 1.3% (**Figure 10**).

A simple empirical formula has been proposed and presented to accurately calculate the lower edge frequency of probe-fed regular hexagonal monopole antennas. The antenna is fed at the vertex of the hexagon and found that the values obtained for lower edge frequency are quite close to the simulated and measured $|S_{11}|$ results of the designed and developed antennas. Lower edge frequency dependency of the probe feeding on the square, circle, and hexagon monopole antenna has been studied using simulation for maximum bandwidth, and the S_{11} results have been demonstrated. The lower edge frequency of the probe-fed hexagonal monopole antenna also depends on the hexagon circle radius, and its variation is also presented. The probe-fed antenna designed using the presented expression demonstrates UWB performance, which ranges from 2.3 GHz to 10.6 GHz with a weak rejection between 5 GHz and 6.5 GHz. The calculated lower and higher edge frequencies of the designed antenna are found to be 2.27 GHz and 11.2 GHz, respectively, which provide an error of only 1.3% and 3.4%, respectively, when compared with the measured S_{11}(dB) results.

6. Gain boosting of a probe-fed hexagonal UWB antenna using an AMC reflector

Due to the existence of higher modes, particularly when fed with a probe, UWB antennas are restricted to having weak gain at higher frequencies. In this section, a technique for increasing the peak gain of a hexagonal UWB antenna fed via probe is presented. An artificial magnetic conductor (AMC)-based reflector is added to the antenna shown in **Figure 11**, enhancing both the peak gain for UWB antennas and the

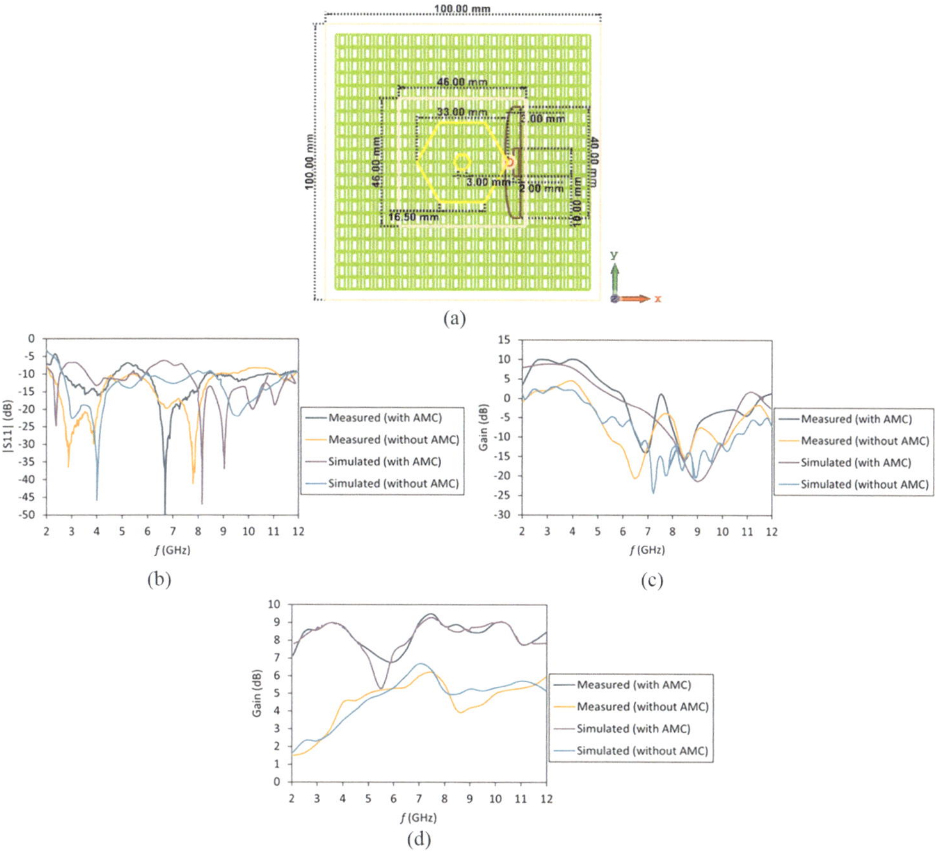

Figure 11.
Probe-fed hexagonal UWB antenna using an AMC reflector: (a) Design; (b) $|S_{11}|$ (dB); (c) boresight gain (dB); (d) peak gain (dB).

boresight gain to a broader band. Peak and antenna boresight gain improvements on average are 3.74 dB and 5.5 dB [39], respectively. In the presence of an AMC-based reflector, the boresight gain increases, becoming positive for a 1 GHz broader band. The UWB antenna measures 46 by 46 mm^2, while the AMC reflector increases the antenna's overall size to 100 by 100 mm^2. The suggested antenna configuration is suitable for UWB applications and may provide a directed and steady radiation pattern.

The presented antenna's boresight and peak gain are compared and shown in **Figure 11c** and **d**, respectively, to help comprehend the gain enhancement phenomena brought on by the AMC reflector. In **Figure 11c**, it is apparent. In the simulation, the AMC reflector increases the average boresight gain by around 5.46 dB. Without an AMC reflector, the antenna's positive boresight gain is visible between 2.2 GHz and 4.3 GHz, whereas with an AMC, a broader band between 2 GHz and 5.8 GHz is seen. The observation of a large increase in boresight gain across a broader band emphasizes the need for an AMC reflector. The boresight gain of the antenna is measured and shown in **Figure 11** to better understand how AMC affects the augmentation of boresight gain. During testing, the antenna's boresight gain seems to be positive across broadband that spans from 2 GHz to 6 GHz as opposed to 2.2 GHz to 4.5 GHz when no

AMC reflector is used. During testing, it is shown that applying AMC to a hexagonal monopole antenna increased the average boresight gain by a factor of 5.5 dB.

The use of an AMC reflector with a square-shaped loop unit cell to convert a monopole-like radiation pattern into a directional pattern increases the gain of the presented antenna. To increase the peak gain and gain of the UWB antenna, an AMC reflector with a 20 by 20 array of unit cells is built at the rear of the hexagonal radiator. After using an AMC reflector, the antenna's boresight and peak gain are both greatly improved by around 5.5 dB and 3.74 dB, respectively. The antenna that has been put together with AMC may be used for UWB applications.

7. Conclusions

Polygonal patch antenna is the keystone of modern wireless communication systems and services. In the development of contemporary wireless communication systems, an antenna with a wide radiation bandwidth is demanded to cater to such a demand. These concerns encourage the antenna researcher to design an antenna with a wide radiation bandwidth.

Hexagon is chosen here to fulfill the requirement of modern wireless communication systems and services while maintaining a fundamental mode. Feed is an important part of the antenna; the impedance and the gain will influence the antenna performance. To increase the antenna's bandwidth, the specified antenna probe feed point is altered. According to the probe feed analysis, a polygonal patch antenna's wideband performance is optimal when the feed point is maintained closer to the polygon's vertex.

Hexagonal patch antennas with probes show narrow band behavior. It is shown how a vertex-fed slotted hexagonal antenna with a truncated half elliptical ground plane responds to SMA connection flanges in terms of impedance bandwidth. The suggested antenna displays C-band characteristics when fed via a connection with a flange, but a flangeless connector may be used to obtain UWB characteristics in a direct-fed antenna. A ground plane reduction approach is also employed that changes the dipole configuration to a monopole configuration to obtain UWB using a hexagonal patch. The ground plane slot enabled the antenna's bandwidth to be increased. Multiple modes in the UWB band are excited using a rectangular slot and a ground plane reduction approach. The antenna has broadband between 2.4 GHz and 10 GHz.

The use of an AMC reflector with a square-shaped loop unit cell to convert a monopole-like radiation pattern into a directional pattern increases the gain of the hexagonal UWB antenna. To increase the peak gain and gain of the UWB antenna, an AMC reflector may be mounted at the rear of a hexagonal radiator. After using an AMC reflector, the presented antenna's boresight and peak gain are greatly improved by around 5.5 dB and 3.74 dB, respectively. The antenna that has been put together with AMC may be used for UWB applications.

Acknowledgements

This research work was funded by the Department of Science and Technology, New Delhi, India (Reference Number: SR/FST/ETI-346/2013) for equipment.

Author details

Abhishek Joshi* and Rahul Singhal
Department of Electrical and Electronics Engineering, Birla Institute of Technology and Science, Pilani, Rajasthan, India

*Address all correspondence to: abhishek.joshi@pilani.bits-pilani.ac.in

References

[1] Pozar DM, Schaubert D. Microstrip Antennas: The Analysis and Design of Microstrip Antennas and Arrays. Piscataway NJ: Institute of Electrical and Electronics Engineers; 1995. pp. 57-104. DOI: 10.1109/9780470545270.ch2

[2] Collins G. Effect of reflecting structures on circularly polarized Tv broadcast transmission. IEEE Transactions on Broadcasting. 1979;**BC-25**:5-13. DOI: 10.1109/TBC.1979.266307

[3] Saleh AAM, Valenzuela R. A statistical model for indoor multipath propagation. IEEE Journal on Selected Areas in Communications. 1987;**5**:128-137. DOI: 10.1109/JSAC.1987.1146527

[4] Lee W, Yeh Y. Polarization diversity system for mobile radio. IEEE Transactions on Communications. 1972;**20**:912-923. DOI: 10.1109/TCOM.1972.1091263

[5] James JR, Hall PS, editors. Handbook of Microstrip Antennas. U.K: Peregrinus; 1989. p. 25-27

[6] Wang YJ, Lee CK, Koh WJ, et al. Design of small and broad-band internal antennas for IMT-2000 mobile handsets. IEEE Transactions on Microwave Theory and Techniques. 2001;**49**:1398-1403. DOI: 10.1109/22.939919

[7] Sabban A. A new broadband stacked two-layer microstrip antenna. In: Proceedings of the 1983 IEEE Antennas Propagation Symposium; 23-26 May 1983; Houston, TX, USA. 1983. pp. 63-66. DOI: 10.1109/APS.1983.1149074

[8] Yang F, Zhang X-X, Ye X, et al. Wide-band E-shaped patch antennas for wireless communications. IEEE Transactions on Antennas and Propagation. 2001;**49**:1094-1100. DOI: 10.1109/8.933489

[9] Tunski Z, Fathy AE, McGee D, Ayers G, Kanamaluru S. Compact multi-band planar antenna for mobile wireless terminals. In: Proceedings of the 2001 IEEE Antennas and Propagation Symposium; 08-13 July 2001; Boston, MA, USA. 2001. pp. 454-457. DOI: 10.1109/APS.2001.959496

[10] Bilotti F, Vegni C. MOM entire domain basis functions for convex polygonal patches. Journal of Electromagnetic Waves and Applications. 2003;**17**:1519-1538. DOI: 10.1163/156939303772681398

[11] Bilotti F, Alù A, Manzini M, Vegni L. Design of polygonal patch antennas with a broad-band behavior via a proper perturbation of conventional rectangular radiators. In: Proceedings of the International Symposium on 2003 IEEE Antennas Propagation; June 2003; Columbus OH. 2003. pp. 268-271. DOI: 10.1109/APS.2003.1219229

[12] Balanis CA. Antenna theory: A review. Proceedings of the IEEE. 1992;**80**:7-23. DOI: 10.1109/5.119564

[13] Kolundzija B, Tasic M, Petrovic N, Mikavica M. Efficient electromagnetic modeling based on automated meshing of polygonal surfaces. In: Proceedings of the International Symposium on Antennas and Propagation Society; 16-21 July 2000. Held in conjunction with: USNC/URSI National Radio Science Meeting (C, Salt Lake City, UT, USA). IEEE; 2000. pp. 2294-2297. DOI: 10.1109/APS.2000.874952

[14] Sorokosz L, Zieniutycz W. On the approximation of the UWB dipole

elliptical arms with stepped-edge polygon. IEEE Antennas and Wireless Propagation Letters. 2012;**11**:636-639. DOI: 10.1109/LAWP.2012.2203575

[15] Lee K-F, Tong K-F. Microstrip patch antennas—Basic characteristics and some recent advances. Proceedings of the IEEE. 2012;**100**:2169-2180. DOI: 10.1109/JPROC.2012.2183829

[16] Chen Y, Wang C-F. From antenna design to feeding design: A review of characteristic modes for radiation problems (invited paper). In: Proceedings of the International Symposium on Antennas & Propagation (ISAP); 23-25 Oct. 2013; Nanjing. China. 2013. pp. 68-70

[17] Garg R. Microstrip Antenna Design Handbook. Boston London: Artech House Publishers; 2001. pp. 9-10

[18] Zhao D, Yang C, Zhu M, et al. Design of WLAN/LTE/UWB antenna with improved pattern uniformity using ground-cooperative radiating structure. IEEE Transactions on Antennas and Propagation. 2016;**64**:271-276. DOI: 10.1109/TAP.2015.2498939

[19] Li D, Mao J. Coplanar waveguide-fed Koch-like sided Sierpinski hexagonal carpet multifractal monopole antenna. IET Microwaves, Antennas & Propagation. 2014;**8**:358-366. DOI: 10.1049/iet-map.2013.0041

[20] Tripathi S, Mohan A, Yadav S. Hexagonal fractal ultra-wideband antenna using Koch geometry with bandwidth enhancement. IET Microwaves, Antennas & Propagation. 2014;**8**:1445-1450. DOI: 10.1049/iet-map.2014.0326

[21] Dastranj A. Low-profile ultra-wideband polarisation diversity antenna with high isolation. IET Microwaves, Antennas & Propagation. 2017;**11**: 1363-1368. DOI: 10.1049/iet-map.2016.0937

[22] Ray KP, Tiwari S. Ultra wideband printed hexagonal monopole antennas. IET Microwaves, Antennas & Propagation. 2010;**4**:437-455. DOI: 10.1049/iet-map.2008.0201

[23] Joshi A, Singhal R. Performance comparison of probe-fed polygonal patch antennas for L-band applications. In: Proceedings of the 2016 IEEE Region 10 Conference (TENCON); 22-25 November 2016; Singapore. 2016. pp. 1307-1311. DOI: 10.1109/TENCON.2016.7848224

[24] Kumar G, Ray KP. Broadband Microstrip Antennas. Boston London: Artech House; 2003. pp. 368-370

[25] Emadeddin A, Salari MA, Zoghi M, et al. A compact ultra-wideband multibeam antenna system. IEEE Transactions on Antennas and Propagation. 2018;**66**:125-131. DOI: 10.1109/TAP.2017.2776342

[26] Thajudeen C, Zhang W, Hoorfar A. Boresight gain enhancement of planar stair-like UWB monopole antenna. Microwave and Optical Technology Letters. 2014;**56**:2809-2812. DOI: 10.1002/mop.28710

[27] Joshi A, Pal D, Singhal R. Vertex-fed hexagram patch wideband antenna. In: Proceedings of the 2014 International Conference on Signal Propagation and Computer Technology (ICSPCT 2014); 12-13 July 2014; Ajmer. 2014. pp. 218-221. DOI: 10.1109/ICSPCT.2014.6885011

[28] Joshi A, Singhal R. Vertex-fed hexagonal antenna with low

cross-polarization levels. Advances in Electrical and Electronic Engineering. 2019;**17**:138-145. DOI: 10.15598/aeee.v17i2.3004

[29] Kishk A, Shafai L. The effect of various parameters of circular microstrip antennas on their radiation efficiency and the mode excitation. IEEE Transactions on Antennas and Propagation. 1986;**34**:969-976. DOI: 10.1109/TAP.1986.1143939

[30] Joshi A, Singhal R. Probe-fed regular hexagonal narrow-slot antenna with reduced ground plane for WLAN applications. In: Proceedings of the 2016 IEEE Region 10 Conference (TENCON); 22-25 November 2016; Singapore. 2016. pp. 1312-1316. DOI: 10.1109/TENCON.2016.7848225

[31] Joshi A, Singhal R. Coaxially fed hexagonal patch antenna for C- and X-band applications with reduced-ground plane. ECTI Transactions on Electrical Engineering, Electronics, and Communications. 2019;**17**:136-143. DOI: 10.37936/ecti-eec.2019172.219184

[32] Malekpoor H, Jam S. Analysis on bandwidth enhancement of compact probe-fed patch antenna with equivalent transmission line model. IET Microwaves, Antennas & Propagation. 2015;**9**:1136-1143. DOI: 10.1049/iet-map.2014.0384

[33] Joshi A, Singhal R. Lower mode excitation in vertex-fed slotted hexagonal S-band antenna. AEU - International Journal of Electronics and Communications. 2018;**87**:180-185. DOI: 10.1016/j.aeue.2018.02.008

[34] Joshi A, Singhal R. Probe-fed hexagonal ultra wideband antenna using flangeless SMA connector. Wireless Personal Communications. 2019;**110**: 973-982. DOI: 10.1007/s11277-019-06768-2

[35] Ray KP, Thakur SS, Deshmukh RA. Broadbanding a printed rectangular monopole antenna. In: Proceedings of the 2009 Applied Electromagnetics Conference (AEMC); Kolkata. 2009. pp. 1-4. DOI: 10.1109/AEMC.2009.5430695

[36] Yang Z, Li L, Wang H. Investigation on ultra-wideband printed circular monopole antenna with frequency-notched. In: 2008 International Conference on Microwave and Millimeter Wave Technology; Nanjing. 2008. pp. 1858-1861. DOI: 10.1109/ICMMT.2008.4540844

[37] Evans JA, Amunann MJ. Planar trapezoidal and pentagonal monopoles with impedance bandwidths in excess of 10:1. In: Proceedings of the IEEE Antennas and Propagation Society International Symposium1999 Digest. Held in conjunction with USNC/URSI National Radio Science Meeting (Cat. No.99CH37010); Orlando FL. 1999. pp. 1558-1561. DOI: 10.1109/APS.1999.788241

[38] Balanis CA. Antenna Theory Analysis and Design. 3nd ed. Wiley; 2008. pp. 816-820

[39] Joshi A, Singhal R. Gain enhancement in probe-fed hexagonal ultra wideband antenna using AMC reflector. Journal of Electromagnetic Waves and Applications. 2019;**33**: 1185-1196. DOI: 10.1080/09205071.2019.1605939

Chapter 3

Wideband True Time Delay Cells

Ahmad Yarahmadi

Abstract

True-time delay (TTD) cells are used in timed array receivers for wideband multi-antenna topologies. TTD cells are divided into two major categories: silicon-based and non-silicon-based structures. Non-silicon-based structures have very good bandwidth but are bulky in the below 10 GHz frequency band. Silicon-based TTD cells are much more compact and better candidates for integrated circuit (IC) design. Passive and active approaches are the two ways to have a silicon-based TTD cell. Passive TTD cells are built by transmission lines (TL), artificial transmission lines (ATL), and LC ladder networks. Their power consumption is very low, and the delay bandwidth is good, but they are still bulky at low frequencies like below 5 GHz applications. Active all-pass filters as TTD cells are presented for these issues. In this chapter, we will discuss the challenges of inductor-based TTD cells. Then, inductor-less TTD cells are presented to address some of the previous structure's issues. Finally, we will talk about these structures' challenges as well. Then, the nonidealities effects on the TTD cell's performance are investigated, and the body bias technique is presented to address these issues.

Keywords: true time delay, inductor-less structure, all-pass filter, timed array structure, inverter-based design, wide band filter

1. Introduction

Nowadays, linearity and dynamic range are critical issues for radars, A/D conversion, multi-standard applications like IoT or 5G/6G communications, receiver chains in communication systems, data processing, and imaging sensors [1–6]. Multi-antenna systems like phased array topologies are suitable for this matter since they can do analog beamforming with very good performance in narrow-band applications [7]. They can do the beamforming task with an approximation of the incident signal. They approximate the delay of the signal with phase change in the phased array multi-antenna system [8]. This approximation is one by phase shifters as the core block of a phased array system. They are quite good for narrow-band signals, but for wideband applications, they suffer from the beam squint phenomena [9].

Unfortunately, other problems arise in wide-band applications like intrinsic narrow-band characteristics, spatial interferences, pulse dispersion, and inter-symbol interference (ISI) [10]. So, another topology must be considered for this matter, leading to the timed array receiver. This structure replaces phase shifters with TTD cells as the system's core in wide-band applications. The TTD cells delay the received signal in contrast to phase shifters which approximate the delay time with a phase shift.

There are two approaches to creating a TTD cell, silicon-based and non-silicon-based TTD cells [2]. Non-silicon-based TTD cells have better performance than silicon-based TTD cells but have some limitations like considerable production cost concerning the rival. So, although the silicon-based TTD cells have a significant loss and high footprint, they are preferable for mass production and low-cost purposes [10]. Silicon-based TTD cells have two separate categories. Passive TTD cells are based on transmission lines (real or artificial) and LC ladders [11, 12]. These topologies have great performance for high frequencies (10 GHz and beyond). But they are very bulky for low frequencies. Active TTD cells are a solution for below 5 GHz applications. in these topologies, all-pass filters are used to create an active TTD cell, and passive inductors are used to ensure the wideband performance of the structure [4, 13].

Figure 1 shows the literature review of TTD cells, and the state-of-the-art design in this area is depicted. This chapter discusses the reason for the high DV (delay-

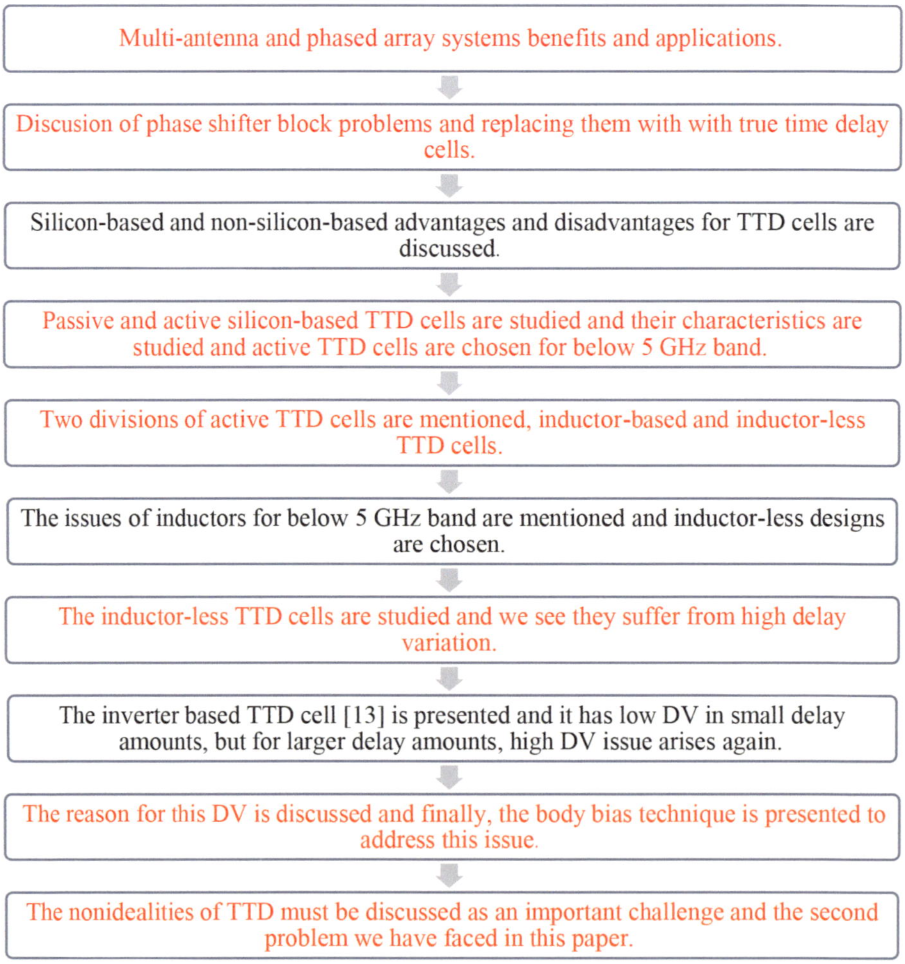

Figure 1.
The diagram of the literature review of TTD cells.

variation) for larger delay amounts in the inverter-based inductor-less active TTD cell. Also, the mechanism of the body biasing scheme for improving the performance of the TTD cell is presented, too.

2. CMOS low-frequency design considerations

Based on the above discussion, an active all-pass filter can be a good solution for designing a true time delay cell. However, there are some design considerations about the low frequency (below 10 GHz) wideband design of TTD cells. In the following paragraphs, we will discuss why an inductor-less TTD cell is needed and why Padé approximation is used to create an all-pass transfer function from the TTD transfer function. Also, we will see the minimum and maximum available delay amounts for a TTD cell with proper characteristics.

2.1 Inductor-less design

In this chapter, we have focused on low-frequency wideband applications. Most of this band's applications demand a very compact structure. Also, the integration ability of circuits is necessary because these cells will be used in the mass production of different devices. So, the chip area and production cost are very important issues, too. On the other hand, the structure must work in various applications. For example, it could be a part of a multi-input multi-output structure like a smartphone, drone, or even an automobile.

An inductor is a bulky and limiting element, and this issue is critical for low frequencies like 1–5 GHz applications because they are large enough in this band. As a result, there is a trade-off here in choosing an inductor [4, 14]. If high-quality off-chip inductors are used, the total cost will increase dramatically. Furthermore, the system's performance will be impacted and bounded if the on-chip inductor is used.

This trade-off could be minimized with advancing technology because on-chip high-quality inductors are available for the advanced chip technologies. Another benefit of these technologies is their smaller parasitic capacitances, which lead us to use inductor-less designs with acceptable wideband performance.

Another issue for inductor-based designs is that the design transformation from one technology node to another (mostly to an advanced one) has a complex procedure. In contrast, it is very simple for an inductor-less design to be transformed into another technology node. This issue is so important concerning the technology improvement speed. Since we almost need to redesign an inductor-based structure, design cost, and total cost are more in these topologies.

However, the chip production cost is higher for more advanced technology nodes. So, there are better choices than bulky elements like inductors below 5 GHz applications. They will impact the total budget of the design dramatically. On the other side, mass production demands production costs as low as possible in most applications.

In the layout and tape-out steps of the chip design, we need to use a trial-and-error procedure to find the proper performance of an inductor-based structure since there is no exact model for them. It forces us to need more design time and lead us to more design cost. Although off-chip inductors are a good candidate to eliminate some of the above issues, they will increase the total cost dramatically, which eventually leads us to a lower yield.

Recently inductor-less active TTD cells have been presented in the literature to address these issues. In these active TTD cells, the transfer function of the cell will be approximated with an all-pass filter, usually with Padé or Taylor approximation. Nevertheless, these topologies suffer from high DV (delay variations) in their wide-band performance [4, 11, 13]. Inverter-based active TTD cells are presented to over-come their DV issue of them in the 1–5 GHz band [15]. This TTD cell delays the incident signal with 10 ps and 3% DV. However, for larger delay values, the DV of the TTD cell will drop dramatically (for example, for a 50 ps delay amount). The body bias technique can be applied to the TTD cell to address this issue.

2.2 Minimum and maximum required delay amount

Delay systems consisted of several delay cells to create a timed array receiver. In a timed array receiver and for a lined antenna topology based on the number of antenna elements, their topology, the distance between the elements, operational frequency, etc. A minimum and maximum required delay can be calculated for a specific purpose [16]. Assume the system has an N-element antenna, the operational frequency is from f_{min} to f_{max}, the maximum steering angle concerning boresight is $\pm\theta$, and b-bit spatial resolution is required. The noise figure (NF) is set to be better than NF_{max}; there is no grating lobe and 40 dB depth for sidelobe is available.

For this preferred system, the distance between antenna elements, maximum delay amount, delay step (minimum delay), and noise figure of each channel can be calculated as follows [17, 18]:

To avoid grating lobes, d, the distance between antenna elements should be less than half of the wavelength of f_{max}:

$$d \leq \frac{\lambda_{f,\max}}{2} \tag{1}$$

The maximum required delay amount of a timed array system is equivalent to the maximum steering angle of the system (θ_{max}). It depends on three terms: the number of antenna elements (N), the distance between antenna elements (d), and the maxi-mum steering angle (θ_{max}). Concerning these terms, the maximum required delay amount of the channel will be [10]:

$$\tau_{\max} = (N-1)\frac{d\sin(\theta_{\max})}{c} \tag{2}$$

In this equation, c is the speed of light in a vacuum. From this equation, and for the incident signal angle concerning bore sigh (θ_{in}), the delay step can be calculated as follows:

$$\Delta\tau = (N-1)\frac{d\sin(\theta_{in})}{c} \tag{3}$$

This delay step in a timed array receiver equals a 4-bit resolution in a phased array receiver system. The minimum delay of the timed array system also can be calculated as the following equation:

$$\tau_{\min} = \frac{\sin(\theta_{\min})}{2f} \tag{4}$$

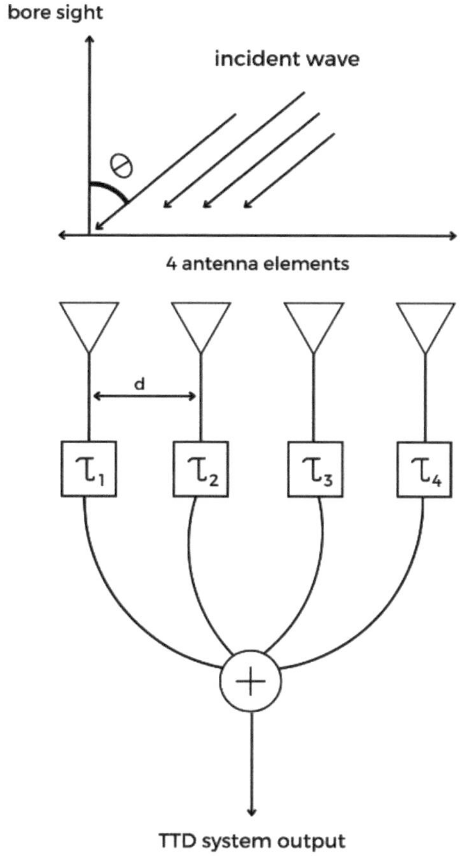

Figure 2.
The timed array system with four antenna elements in the lined topology.

Based on the above discussion, four antenna elements in a lined topology are required for an assumed 1–5 GHz application and 60 degrees of steering angle with a 4-bit resolution. The distance between them will be d = 3 cm, the maximum delay amount will be 260 ps, and the minimum delay amount will be 12.5 ps.

With this information, we can design the required delay system as shown below, so the delay block, which means a true time delay cell, can be designed for this system (**Figure 2**).

2.3 Delay approximation accuracy analysis

The two most common approaches to approximating true time delay transfer function (*TF*) are the Padé approximation and Taylor approximation. The Taylor expansion of ideal delay *TF* can be written as follows [19]:

$$e^{-sT} \simeq 1 - \frac{(sT)}{1!} + \frac{sT^2}{2!} + \ldots + (-1)^n \frac{sT^n}{n!} \tag{5}$$

However, this *TF* does not meet the physical requirements to create a filter. To make it feasible for filter design, we must modify this equation with the following one:

Approximation order	Transfer function
1	$\frac{2-(sT)}{2+(sT)}$
2	$\frac{8-4(sT)+(sT)^2}{8+4(sT)+(sT)^2}$
3	$\frac{48-24(sT)+6(sT)^2-(sT)^3}{48+24(sT)+6(sT)^2+(sT)^3}$
4	$\frac{384-192(sT)+48(sT)^2-8(sT)^3+(sT)^4}{384+192(sT)+48(sT)^2+8(sT)^3+(sT)^4}$
5	$\frac{3840-1920(sT)+480(sT)^2-80(sT)^3+10(sT)^4-(sT)^5}{3840+1920(sT)+480(sT)^2+80(sT)^3+10(sT)^4+(sT)^5}$

Table 1.
The Taylor approximation for the time delay transfer function.

$$e^{-sT} = e^{-s\left(\frac{T}{2}+\frac{T}{2}\right)} = e^{-s\frac{T}{2}}e^{-s\frac{T}{2}} = \frac{e^{-s\frac{T}{2}}}{e^{s\frac{T}{2}}} \tag{6}$$

As a result, this modification improves the approximation's accuracy, so the Taylor approximation is applicable for delay approximation for below 4th-order approximations. The fifth order and above approximations will face instability issues and are not available for filter design. The Taylor approximation for 1 to 5-order approximations is shown in **Table 1**.

Based on these five transfer functions, the MATLAB simulation for their step response concerning true time delay is done, and the results as the order of approximation increases, the precision increases, too.

Another approach to creating the *TF* of an all-pass filter is the Padé approximation. In this case, the *TF* of true timed delay will approximate with this eq. [20]:

$$e^{-sT} \simeq \sum_{i=0}^{n+m} (-1)^i \frac{sT^i}{i!} = \frac{\sum_{i=0}^{m} p_i (sT)^i}{\sum_{i=0}^{n} q_i (sT)^i} \tag{7}$$

Table 2 shows the *TF* of the approximation for one to five orders. Furthermore, **Figure 3** depicts their frequency response versus the ideal step response. It is worth mentioning that in practical designs, because of unstable zeros of *TF*, below 10-order approximations are feasible. This approximation does not need any modifications for lower orders. So, it can be a better choice for simple designs. Based on these five transfer functions, the MATLAB simulation for step response of them concerning true time delay is done, and the results are shown in **Figure 3**.

The simulations of both approximations show that the Padé approximation has a more accurate response concerning Taylor approximation. For example, for the 5-th-order approximation, the accuracy of Padé is 30 times better than Taylor's approximation.

So, the first step to designing a TTD cell is to use a Padé approximation with proper order for an intended purpose. Because lower orders are simple, but higher orders are

Approximation order	Transfer function
1	$\frac{2-(sT)}{2+(sT)}$
2	$\frac{12-6(sT)+(sT)^2}{12+6(sT)+(sT)^2}$
3	$\frac{120-60(sT)+12(sT)^2-(sT)^3}{120+60(sT)+12(sT)^2+(sT)^3}$
4	$\frac{1680-840(sT)+180(sT)^2-20(sT)^3+(sT)^4}{1680+840(sT)+180(sT)^2+20(sT)^3+(sT)^4}$
5	$\frac{30240-15120(sT)+3360(sT)^2-420(sT)^3+30(sT)^4-(sT)^5}{30240+15120(sT)+3360(sT)^2+420(sT)^3+30(sT)^4+(sT)^5}$

Table 2.
The Padé approximation for the time delay transfer function.

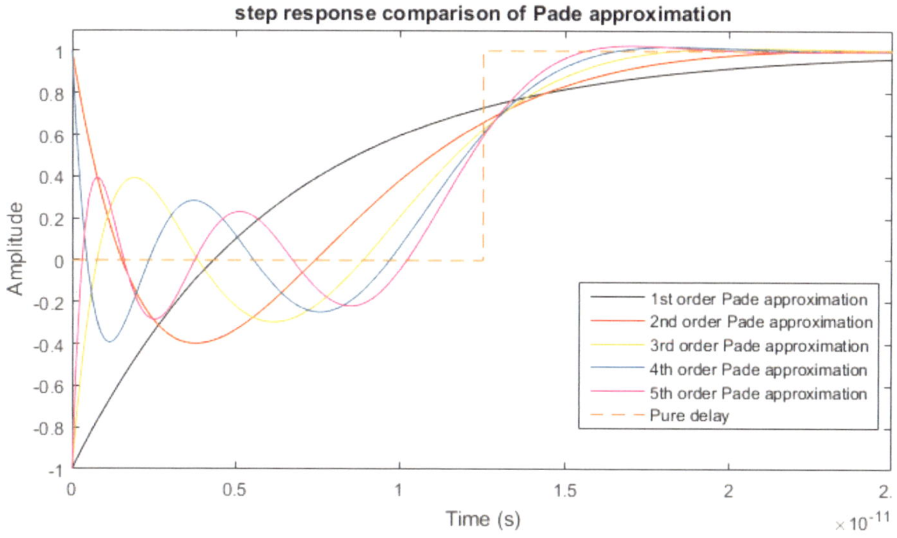

Figure 3.
The step response of equal order numerator and denominator of the TF's fraction for Padé approximation versus TTD for a 5-second delay.

more accurate. For our TTD cell, we have focused on simplicity, So the lower orders are chosen.

Another important issue that is remained here is the difference between true time delay (τ_d) and group delay (τ_g). To investigate more about them, first, the phase response of both first and second-order approximations are calculated here:

$$\varphi_{1st,order}(\omega) = -2\tan^{-1}\frac{\omega T}{2} \quad (8)$$

$$\varphi_{2nd,order}(\omega) = -2\tan^{-1}\frac{6\omega T}{12-\omega^2T^2} \quad (9)$$

From these two equations, the delay amount of both approximations can be calculated:

$$\tau_{d1}(\omega) = -\frac{\varphi_{1st,order}(\omega)}{\omega} = \frac{2tan^{-1}\frac{\omega T}{2}}{2} \tag{10}$$

$$\tau_2(\omega) = -\frac{\varphi_{2nd,order}(\omega)}{\omega} = \frac{2tan^{-1}\frac{6\omega T}{12-\omega^2T^2}}{2} \tag{11}$$

Furthermore, from these equations, the group delay of both approximations can be calculated:

$$\tau_{g1}(\omega) = -\frac{d}{d\omega}\left(\varphi_{1st,order}(\omega)\right) = \frac{2}{1+\frac{\omega T}{2}} \tag{12}$$

$$\tau_{g2}(\omega) = -\frac{d}{d\omega}\left(\varphi_{2nd,order}(\omega)\right) = \frac{2\left(73+\omega^3T^3\right)}{\left(12+35\omega^2T^2\right)\left(12-\omega^2T^2\right)} \tag{13}$$

2.4 Maximum available flat delay amount

From these four equations, the group delay and true time delay are equal when the phase changes are linear, i.e., linear phase response. This matter can be calculated from these equations:

$$\begin{aligned}
&\tau_g = \tau_d \\
&\Rightarrow -\frac{\varphi(\omega)}{\omega} = -\frac{d\varphi(\omega)}{d\omega} \\
&\Rightarrow \frac{d\omega}{\omega} = \frac{d\varphi(\omega)}{\varphi(\omega)} \\
&\Rightarrow \int\frac{d\omega}{\omega} = \int\frac{d\varphi(\omega)}{\varphi(\omega)} \\
&\Rightarrow ln\left(\varphi(\omega)\right) = ln\left(\omega\right) + c \\
&\Rightarrow \varphi(\omega) = e^{ln\left(\omega\right)+c} = e^{ln\left(\omega\right)}e^{c} \\
&\Rightarrow \varphi(\omega) = C.\omega
\end{aligned} \tag{14}$$

From the above calculations, and concerning the value of phase, we can write the below equation for both first and second-order approximations:

$$\varphi(\omega) = -2Arctan\left(\frac{\omega T}{2}\right) \overset{Linear}{=} C\omega \tag{15}$$

$$\varphi(\omega) = -2Arctan\left(\frac{6\omega T}{12-(\omega T)^2}\right) \overset{Linear}{=} C\omega \tag{16}$$

Both equations are arctangent-based. As shown in **Figures 4–6**, the response will be linear up to 23 degrees (0.4 radians) which means, until the phase argument reaches this amount, the output of the TTD cell will remain linear.

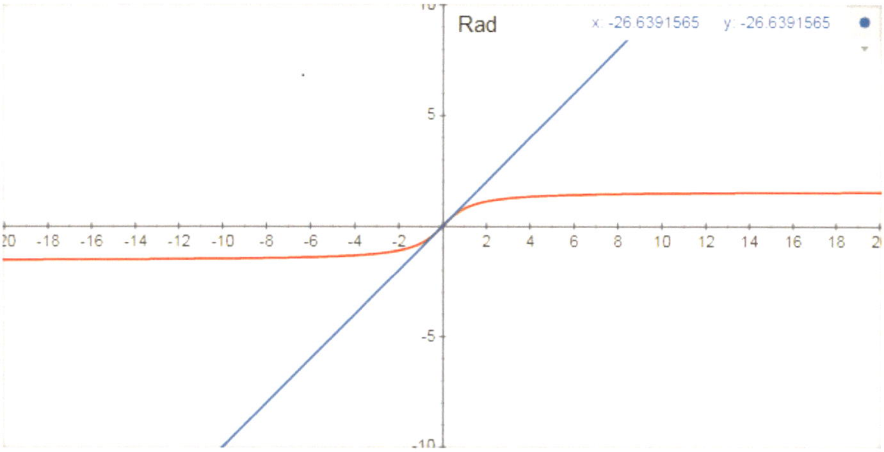

Figure 4.
The comparison between y = arctan (x) *and* y = x.

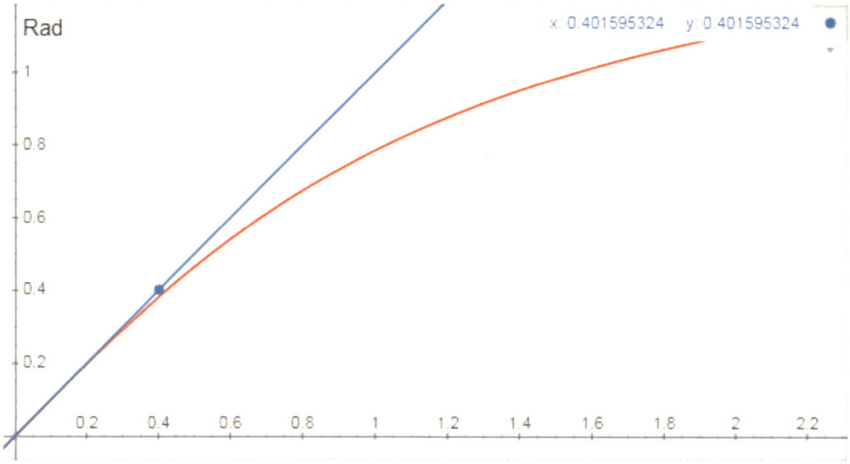

Figure 5.
The comparison between y = arctan (x) *and* y = x *in magnified mode.*

So, for the 5 GHz application, this maximum delay can be calculated from the previous Eqs. (15) and (16). For the first-order approximation, the delay amount will be T = 25 ps, and for the second-order approximation, the delay amount will be T = 24 ps. Up to this amount, the TTD cell will remain in a linear area, and the group delay and true time delay are equal.

It is why most active TTD cells with larger delay values suffer from DV in lower frequencies. For example, [13, 14] tried to provide 24 ps and 59 ps delays, but they have 30% and 50% DV. [11, 21] also have a borderline 10% DV for their bandwidth.

So, for having larger delays, it is inevitable to use circuit-level or system-level approaches like master-save, DLL (delay lock loop), and phase linearizer to address this issue [10, 22, 23].

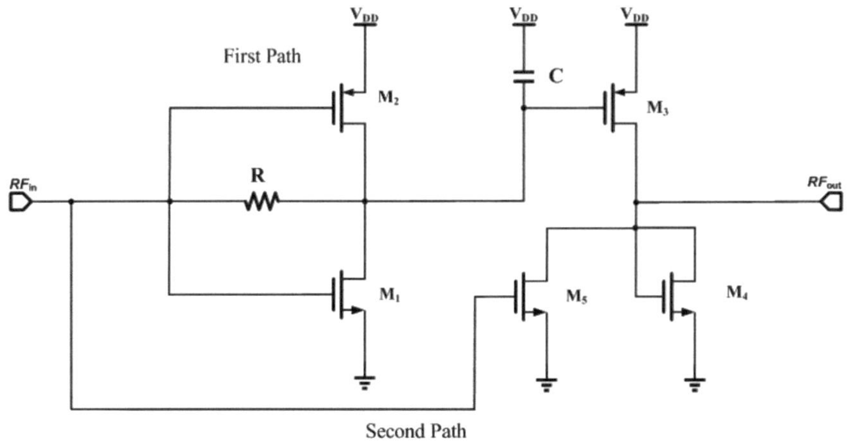

Figure 6.
The inverter-based inductor-less TTD cell is proposed in [15].

2.5 Inductor-less TTD cell with COMS inverter cell

As mentioned above, the ideal transfer function of a TTD cell can be approximated with Padé approximation [15]. The equation of this approximation can be written as follows:

$$H(s) = e^{-s\tau} \approx \frac{2 - s\tau}{2 + s\tau} = \frac{1 - s\left(\frac{\tau}{2}\right)}{1 + s\left(\frac{\tau}{2}\right)} = -\left(\frac{-2}{1 + s\left(\frac{\tau}{2}\right)} + 1\right) \tag{17}$$

This equation shows that the all-pass filter can be created with a low pass filter and a gain stage. The inverter-based structure for a TTD cell is proposed in [15] as shown in **Figure 6**.

In this TTD cell, the second path is the gain stage with a gain of 2. It consists of M_4 and M_5. The first pass is the low pass filter section which consists of M_1, M_2, M_3, R, and C. To investigate the structure's performance, first, we must calculate the low-frequency gain of the cell. The overall gain of the structure ignoring parasitic capacitances comes from this equation:

$$\frac{v_{out}}{v_{in}} = A_{v1} + A_{v2}$$

$$A_{v1} = \frac{\frac{g_{m3}}{g_{m4}+1/R_L}\left(1 - R\left(g_{m1} + g_{m2}\right)\right)}{1 + sCR}, A_{v2} = \frac{g_{m5}}{g_{m4} + 1/R_L} \tag{18}$$

In this equation, A_{v1} and A_{v2} are the low-frequency gains of the first and second stages. The overall gain is the summation of these two gains since the two paths are parallel. In this equation, R and C are the feedback resistor and second-stage capacitances of the TTD cell. R_L is the output load (usually 50 ohms). $g_{m1}, g_{m2}, g_{m3}, g_{m4}$, and g_{m5} are the M_1 to M_5 transconductances. This gain can be rewritten as follows if we made some assumptions:

$$\frac{v_{out}}{v_{in}} = A_{v1} + A_{v2} = \frac{1 - sRC}{1 + sRC}$$

if

$$g_{m5} = g_{m4} + 1/R_L, \frac{g_{m3}(1 - R(g_{m1} + g_{m2}))}{g_{m4} + 1/R_L} = -2 \tag{19}$$

With this assumption, the all-pass filter with equal frequency for the left-handed pole and right-handed zero is created. From the previous discussion, the true time delay of this TTD cell can be calculated as follows:

$$\tau_d(\omega) = \frac{2\tan^{-1}\omega RC}{\omega} \tag{20}$$

This equation means that the delay of the TTD cell can be manipulated just with the R and C of the TTD cell. Although all parasitic parameters can change the value, R and C are the dominant parameters for this matter. While R is used as the feedback resistor and ensures the wideband application of the structure, capacitor C is relaxed to create an arbitrary delay amount. There are many trade-offs and non-idealities in the TTD cell design, like input capacitance of the stage, parasitic zero and pole, limited bandwidth, finite output impedance, noise, and nonlinearity of the structure. They are deeply discussed in [15]. The TTD cell must have input/output impedance matching since it may work in a delay line.

From **Figure** 7, the TTD cell provides a 10.6 ps delay amount with 3% DV in the 1–5 GHz. The loss of TTD cells is acceptable, and S_{21} is better than −3.9 dB. The TTD cell has input/output impedance matching, and S_{11} and S_{22} are below −10 dB. This TTD cell works well to provide a 10 ps delay amount. However, a larger delay will have a bounded delay bandwidth, and the DV will arise, as discussed earlier. Another issue for this novel structure is its performance against process variations, aging effects, field variations, and other non-idealities. We need a solution to ensure its performance against these issues.

2.6 The TTD cell improvement

There are two major issues in inductor-less TTD cells. The High value of DV for large delay amounts and non-ideality issues like process, supply voltage, and temperature (PVT) variations, aging effects, and mismatch between TTD cell's devices. A few approaches to address these issues are discussed in the introduction section. Here we will discuss another approach to improve the TTD cell's performance.

As the semiconductor industry develops, the size of the transistors decreases. This size reduction causes some stability issues for transistors, especially for high-frequency applications. Due to this size reduction, the transistors are more sensitive to PVT variations, aging effects, and field variations.

The effect of these non-idealities will show on the change in threshold voltage of the transistor, which changes the transistor's transconductance and noise parameters of the device. The threshold voltage can be defined as a function of three parameters as follows:

$$V_t = \mathrm{V}_{t0} + \gamma\left(\sqrt{(|2\Phi_F| - \mathrm{V_{BS}})} - \sqrt{|2\Phi_F|}\right) \tag{21}$$

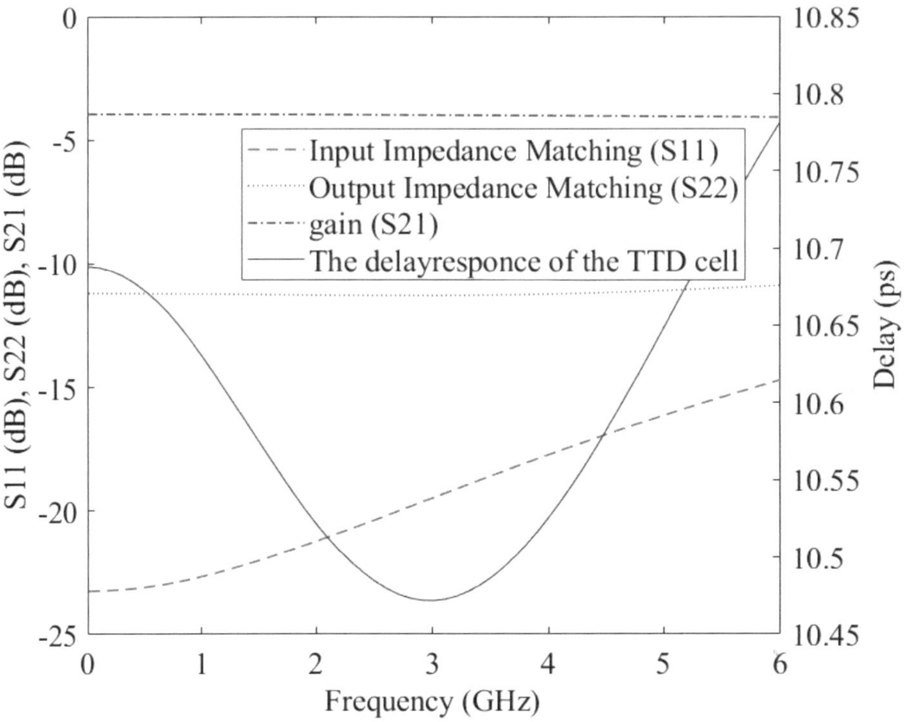

Figure 7.
The S_{11}, S_{22}, gain, and delay of the TTD cell.

In this equation, V_t is the threshold voltage when the body bias is available (V_{BS}, not zero), V_{t0} is the threshold voltage when the body bias (V_{BS} = 0 V) is zero, $\Phi_?$ is the Fermi potential, and usually is 0.3–0.4 $V^{1/2}$. γ is the body parameter, also known as the body coefficient, and comes from the below equation:

$$\gamma = \frac{t_{ox}}{\varepsilon_{ox}}\left(\sqrt{2q\varepsilon_{Si}NA}\right) \tag{22}$$

In this equation, t_{ox} is oxide thickness, ε_{ox} is the permittivity of oxide,ε_{Si} is the permittivity of silicon, NA is the doping density, and q equals the electrical charge. γ is always positive.

From Eq. (21), Process variations change the body coefficient, which leads to V_t variation. Temperature variations and field effects change the Fermi potential level, which leads to V_t variation, too. Fortunately, from Eq. (21), another phenomenon can be observed. If the threshold voltage faces any change by the non-idealities, we can manipulate it by V_{BS} to compensate for that variation deliberately.

With this technique, we can provide a larger delay with the proposed TTD cell since any delay variation can be tuned by the body bias of the TTD cell's devices. Moreover, the structure will be robust against any non-ideality from PVT, again effects or field effects.

The results of the body bias technique are depicted in **Figure 8**. Delay, S_{11}, S_{22}, and S_{21} with and without body bias techniques are depicted in this figure. The delay

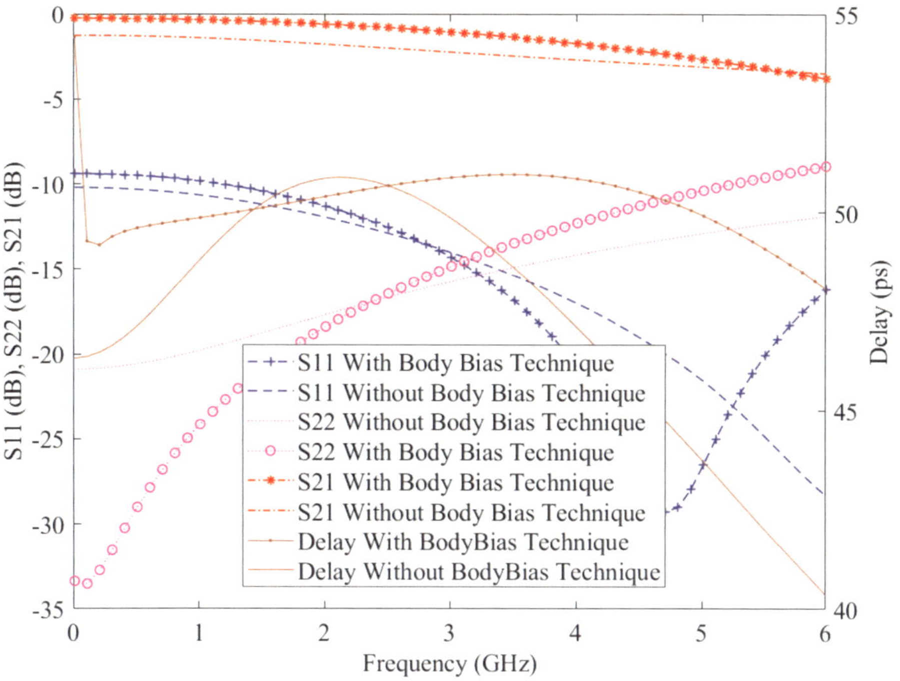

Figure 8.
The Results of TTD cell with and without body bias technique.

amount of the TTD cell is 50.95 ps with only 2% DV in the 1–5 GHz band. The TTD shows a flat delay response compared to the TTD without body bias. The loss (S_{21}) of the TTD cell is improved, and we have a flatter S_{21} parameter. S_{11} and S_{22} are below −10 dB and improved with the body bias technique.

3. Conclusion

In this chapter, we have discussed wide-band TTD cells. There are two major TTD cells, silicon-based and non-silicon-based TTD cells. In this chapter, we studied why silicon-based TTD cells are used in today's IC design. Then we faced two major approaches for designing a silicon-based TTD cell. Active TTD cells are used for high frequencies (upper than 10 GHz), and active TTD cells are used for low frequencies (below 5 GHz). Active TTD cells have two topologies, inductor-based TTD cells, and inductor-less TTD cells. This chapter deeply discussed why we do not prefer inductors in our designs. Also, active TTD cells suffer from high DV, which its roots are discussed here. Finally, we have presented an inductor-less inverter-based TTD cell. This cell works well, but if we need large delay amounts, it will face high DV, too. The body bias technique is applied to the TTD cell to overcome the DV issue. The results are fruitful, and the structure provides larger amounts of delay with a flat frequency response. Moreover, the TTD cell is robust against non-idealities like PVT variations, aging effects, and field effects.

Author details

Ahmad Yarahmadi
Tarbiat Modares University, Tehran, Iran

*Address all correspondence to: ahmad.yarahmadi@modares.ac.ir

References

[1] Jeong J, Collins N, Flynn MP. A 260 MHz IF sampling bit-stream processing digital beamformer with an integrated array of continuous-time band-pass ΔΣ modulators. IEEE Journal of Solid-State Circuits. 2016;**51**(5):1168-1176

[2] Elamien MB, Maundy BJ, Belostotski L, Elwakil AS. Ultra-low-power compact single-transistor all-pass filter with tunable delay capability. AEU-International Journal of Electronics and Communications. 2021;**132**:153645

[3] Spoof K, Unnikrishnan V, Zahra M, Stadius K, Kosunen M, Ryynänen J. True-time-delay beamforming receiver with RF re-sampling. IEEE Transactions on Circuits and Systems I: Regular Papers. 2020;**67**(12):4457-4469

[4] Paul A, Ramírez-Angulo J, Lopez-Martin AJ, Carvajal RG. CMOS first-order all-pass filter with 2-Hz pole frequency. IEEE Transactions on Very Large Scale Integration (VLSI) Systems. 2019;**27**(2):294-303

[5] Guo B, Wang H, Wang H, Li L, Zhou W, Jalali K. A 1–5 GHz 22 mW receiver frontend with active-feedback baseband and voltage-commutating mixers in 65 nm CMOS. IET Circuits, Devices & Systems. 2022;**16**(7):543-552

[6] Nejadhasan S, Zaheri F, Abiri E, Salehi MR. PVT-compensated low-voltage and low-power CMOS LNA for IoT applications. International Journal of RF and Microwave Computer-Aided Engineering. 2020;**30**(11):e22419

[7] Mondal I, Krishnapura N. A 2-GHz bandwidth, 0.25–1.7 ns true-time-delay element using a variable-order all-pass filter architecture in 0.13 μm CMOS. IEEE Journal of Solid-State Circuits. 2017;**52**(8):2180-2193

[8] Spoof K, Zahra M, Unnikrishnan V, Stadius K, Kosunen M, Ryynänen J. A 0.6–4.0 GHz RF-resampling beamforming receiver with frequency-scaling true-time-delays up to three carrier cycles. IEEE Solid-State Circuits Letters. 2020;**3**:234-237

[9] Gao F, Wang B, Xing C, An J, Li GY. Wideband beamforming for hybrid massive MIMO terahertz communications. IEEE Journal on Selected Areas in Communications. 2021;**39**(6):1725-1740

[10] Garakoui SK, Klumperink EAM, Nauta B, van Vliet FE. Compact cascadable gm-C all-pass true time delay cell with reduced delay variation over frequency. IEEE Journal of Solid-state Circuits. 2015;**50**(3):693-703

[11] Chen Y, Li W. An ultra-wideband pico-second true-time-delay circuit with differential tunable active inductor. Analog Integrated Circuits and Signal Processing. 2017;**91**(1):9-19

[12] Kim J, Park J, Kim J-G. CMOS true-time delay IC for wideband phased-array antenna. ETRI Journal. 2018;**40**(6): 693-698

[13] Aghazadeh SR, Martinez H, Saberkari A. 5GHz CMOS all-pass filter-based true time delay cell. Electronics. 2018;**8**(1):16

[14] Aghazadeh SR, Martinez H, Saberkari A, Alarcon E. Tunable active inductor-based second-order all-pass filter as a time delay cell for multi-GHz operation. Circuits, Systems, and Signal Processing. 2019;**38**(8):3644-3660

[15] Yarahmadi A, Jannesari A. Wideband inductorless true time delay cell based on CMOS inverter for timed

array receivers. Circuits, Systems, and Signal Processing. 2021;**40**(8):3703-3726

[16] Cho M-K, Song I, Kim J-G, Cressler JD. An active bi-directional SiGe DPDT switch with multi-octave bandwidth. IEEE Microwave and Wireless Components Letters. 2016; **26**(4):279-281

[17] Visser HJ. Array and Phased Array Antenna Basics. England: John Wiley & Sons; Chichester; 2005

[18] De Oliveira AM, Perotoni MB, Garay JRB, Barboza SHI, Justo JF, Kofuji ST. A complete CMOS UWB Timed-Array Transmitter with a 3D vivaldi antenna array for electronic high-resolution beam spatial scanning. In: 2013 SBMO/IEEE MTT-S International Microwave & Optoelectronics Conference (IMOC), 4–7 August 2013. Rio de Janeiro, Brazil: IEEE; 2013. pp. 1-6

[19] Makundi M, Valimaki V, Laakso TI. Closed-form design of tunable fractional-delay allpass filter structures. In: ISCAS 2001, the 2001 IEEE International Symposium on Circuits and Systems, 6–4, 2001. Vol. 4. Sidney, Australia: IEEE; 2001. pp. 434-437

[20] Kashmiri SM, Haddad SAP, Serdijn WA. High-performance analog delays: surpassing Bessel-Thomson by Pade-approximated Gaussians. In: 2006 IEEE International Symposium on Circuits and Systems, 21–24 May 2006. Kos, Greece: IEEE; 2006. pp. 2349-2352

[21] Chang Y-W, Yan T-C, Kuo C-N. Wideband time-delay circuit. In: 2011 6th European Microwave Integrated Circuit Conference, 10–11 October 2021. Manchester, UK: IEEE; 2021. pp. 454-457

[22] Chen Y, Li W. Compact and broadband variable true-time delay line with DLL-based delay-time control. Circuits, Systems, and Signal Processing. 2018;**37**(3):1007-1027

[23] Nauta B. Analog CMOS Filters for Very High Frequencies. New York, USA: Springer; 1993

Section 2

UWB Technology-Based Applications

Chapter 4

Ranging and Positioning with UWB

Jerome Henry

Abstract

Indoor location is one of the key use cases enabled by UWB for navigation and asset tracking. The 802.15.4a and 802.15.4z standards describe several techniques for determining the distance between a mobile device client and a set of static anchors. Two of them (SS-TWR and DS-TWR) use a bidirectional exchange between the client and the anchor, with the resulting distance being calculated on the initiating side (SS-TWR) or both sides (DS-TWR). OWR does not require an exchange and simply relies on the comparison of arrival times of signals (TDoA). With UL-TDoA, the time of arrival of the client signal is compared on several anchors, drawing distance hyperbolae from which the client location is deduced. With DL-TDoA, the reverse happens: the client compares the time of arrival of signals from several anchors and deduces its position. A third family of techniques is not described in the Standard but is commonly implemented in the field: AoA, where a comparison of the phase of the signal among two or more antennas is used to compute the direction of the sender. From these elements, a location engine computes the mobile device's position. This chapter examines these techniques in detail.

Keywords: UWB, 802.15.4a, 802.15.4z, TWR, OWR, TDoA, SS-TWR, SDS-TWR, DS-TWR, PRF, RDEV, ERDEV, ToF, RRMC IE, RRTI IE, RMI IE

1. Introduction

It is often said that outdoor localization is on its way to being solved, thanks to the progress of the fusion of GPS, dead reckoning, and cellular techniques. Indoor, however, the challenge remains. Although they vary widely, all indoor localization uses cases revolve around the idea of determining the position of a known object about known landmarks, like room numbers or other locally significant markers. Sensing techniques allow for the detection of a moving body, but the accuracy of such determination is limited. Precision can greatly increase if the object incorporates a radio frequency (RF) technology, that allows it to interact with other RF objects whose location is known. These conditions bring the problem closer to its outdoor counterpart, and many technologies attempt to solve it.

Among them, Ultra-Wide Band (UWB) technologies have emerged as solutions of choice. "Ultra-wide" is a term characterizing any radio transmission occurring over a large channel (> 500 MHz). In most cases, these technologies face the risk of interfering with other transmitters and are allowed to transmit this type of large signal in exchange for the signal to be very low power (thus limiting the interferences they cause to others).

 IntechOpen

There have been many proposals for UWB communications, in many forums. However, one of them, first defined in the IEEE 802.15.4a Standard in 2007 [1], then refined in the IEEE 802.15.4z Standard in 2020 [2], has emerged as a key player for indoor localization, because of its claim of high accuracy. Experiments in controlled environments report localization accuracy down to 3 cm (1 inch), and commercial deployments now claim less than 30 cm (1 foot) error. This precision is made possible by the characteristics of the UWB transmission, as defined in the IEEE 802.15.4 family of standards, and augmented by industry-wide certifications like FiRa. This chapter examines the principles and components that guide UWB-ranging and make it the de facto solution for indoor ultra-precise localization.

2. Ranging with UWB

2.1 UWB ranging claim to accuracy

UWB may appear as just one of many radio frequency-based (RF) technologies interested in the localization use case. However, its design makes it particularly well-adapted for accurate ranging. Most radio technologies have attempted, in one way or another, to measure the distance between a sender and a receiver but have found the endeavor to be challenging.

A simple approach is to translate the received signal strength value into a distance estimation, using standard free path loss equations. One obvious limitation of such a technique is that obstacles may cause the received signal to be weaker than it would be, traveling in an unobstructed path. Reflections and multipath in general may also cause the received signal to be stronger (constructive interferences) or weaker (destructive interferences) than it would be in an open space. For these reasons, signal strength-based techniques for distance estimation are used, but not preferred, as they have the reputation of being flawed with large inaccuracy.

Other techniques measure the angle of the received signal on multiple receivers (or the angle from multiple transmitters, which positions are known). By using these combined angles, the transmitter or the receiver location can be deduced with standard geometrical tools. UWB allows for this technique as we will see later in this chapter, and its accuracy compares to that of other technologies. One requirement of this approach is the cooperation between several senders or several receivers to construct a geometrical object before the location can be found (while the signal strength-based techniques directly translate the signal at a single receiver into a distance).

A third family of techniques attempts to measure the time-of-flight (ToF), that is the time taken by a signal to travel between the transmitter and the receiver. This approach has received the favor of many radio technology families, including IEEE 802.11 Fine Timing Measurement (FTM), Bluetooth Low Energy (BLE) High-Accuracy-Distance-Measurement (HADM), and UWB, because it can lead to very accurate results. However, it requires the protocol designers and implementers to solve several technical difficulties. In addition to the challenge of agreeing on a common time reference between the transmitter and the receiver, which we will examine further in the next section, ToF requires that the receiver should be able to precisely determine the time of arrival of the signal. For UWB, just like for most other techniques, this time is the precise time of arrival of the beginning of the signal, often referred to as the first pulse of the first symbol of the header of a PPDU (Physical Protocol Data Unit). The PPDU includes the physical layer of the transmitted frame, which typically starts with a form of the preamble (called the SYNC field in 802.15.4), a simple rhythmic structure that precedes the body of the frame, where the real data will be found. The preamble serves three core purposes:

- In an environment where some form of RF noise is always expected, the preamble allows the receiver to recognize that the amount of energy reaching the antenna is strong enough that it is likely that some form of intentional transmission is occurring (and thus the receiver should attempt to measure that energy to recognize the transmission)

- Even if the receiver has missed the very beginning, the known structure of the preamble soon allows the receiver to catch up, recognizing the preamble for what it is, and deducing the part it missed from the parts it recognizes

- The preamble indicates how the rest of the frame should be interpreted (how it is encoded or modulated).

Determining the exact time of the arrival of the first part of the preamble is more challenging than it would seem to the untrained eye. The receiver needs to measure the energy over a range of interesting frequencies (the channel to which the receiver is set) at regular intervals. To measure the rise and fall of energy of a signal that follows the structure of a sinusoid wave, the sampling typically needs to occur (at least) twice as fast as the channel bandwidth. For example, if the channel is 1 Hertz wide (one peak and one trough per second), measuring the energy twice per second is sufficient. Similarly, if the channel is 80 MHz wide (e.g., with IEEE 802.11 ac), sampling the channel 160 million times per second is needed. This second case represents taking a sample once every 6.25 nanoseconds (ns). Practically, this method means that the receiver takes a sample at a time t_0 and measures no notable energy beyond the noise floor. Then at time t_1, 6.25 ns later, another sample is taken, and this time significant energy is detected (and thus a preamble is likely arriving at the antenna). However, the preamble may have started arriving at any time between t_0 and t_1. Light and any other RF energies travel at approximately 299,792,458 m per second in the air and therefore travel a bit more than 1.80 m (or 6 ft) in a 6.25 ns interval. This sampling method makes it would be difficult to measure a distance with a precision greater than 1.80 m over our example 80 MHz-wide channel. Technologies leveraging channels mechanically obtain better-ranging accuracy with the ToF technique, because they sample more often, reducing the time of arrival uncertainty. UWB commonly uses ultra-wide channels, 500 MHz or wider, and thus benefits from a clear advantage in this domain.

Technologies that focus on high data rates need to implement a rich modulation structure, where many symbols are sent in parallel over the width of the channel. The advantage is data transmission speed, but the downside is another challenge for efficient ToF. To receive the data part of the signal properly, the receiver needs to measure the energy of each segment of interest (where each symbol may be found) over the entire channel width. However, segments with a strong energy peak may overwhelm neighboring segments with low energy. One important requirement for these types of technology is therefore to mandate a low Peak to Average Power Ratio (PAPR), to ensure that no segment blinds its neighbors by being much over or below the average energy of all segments taken together. This necessity makes that many of these complex modulations allow for sidelobes of energy on each segment, that smoothen the average energy value of the segment. Unfortunately, a side effect of this structure is that the sidelobe may reach the receiver before the useful part of the segment. This event is inconsequential for proper reception of the data associated with the segment (the receiver can recognize the peak and demodulate its associated symbol), but it may lure the receiver into believing that the signal started being received before it was, causing the system to conclude on unrealistically short distances. ToF is challenging for these modulation-rich technologies.

UWB was designed to avoid both the sidelobe and the bandwidth issues. In the time domain, the UWB signal is composed of very short pulses (2 ns each). The sequence of pulses encodes the message to transmit. The interval between each pulse (usually represented by the term pulse repetition frequency [PRF]) determines how much data can be sent by the unit of time (**Figure 1**). Because each pulse is very short, it can easily be recognized (no issue of sidelobe confusing the receiver). Because the receiver recognizes the pattern of the preamble, it can recompose the first pulse (and its exact arrival time) even if the receiver did not sample perfectly at the right point in time. This structure gives UWB the claim of a range accuracy of the order of 2 inches, or 6 cm [3]. We will see below that there are still some technical issues to solve to get to that level.

Another advantage of the UWB transmission structure is that the pulse is sent, in the frequency domain, over the large 500-MHz channel. The amount of energy transmitted is large enough that the receiver can recognize each pulse. However, the amount of energy per unit of bandwidth is very small (e.g., in the domains regulated under the American Federal Communications Commission [FCC], −41.3 dBm per MHz maximum; by contrast, some channels for 802.11 transmissions allow 11 dBm/MHz, i.e., a signal one hundred thousand times more powerful per MHz of bandwidth than UWB). This spread of the energy transmitted makes those narrower systems (for example, an 802.11 device listening to "only" 80 MHz of the 500 MHz-wide transmissions) will barely detect the UWB signal in the general noise. However, the UWB receiver, capturing the full 500 MHz, will read each pulse with ease.

2.2 Single-sided two-way ranging

2.2.1 Ranging terminology

The UWB system of short pulses allows for good-ranging precision. The UWB ranging frame itself is not very special. What distinguishes it from any other UWB frame is that one bit of the Physical (PHY) Layer (the ranging bit) is set. The rest of the header, and the rest of the frame, can be of any of the formats allowed by the 802.15.4a or 802.15.4z Standards. If the frame is solely intended for ranging purposes,

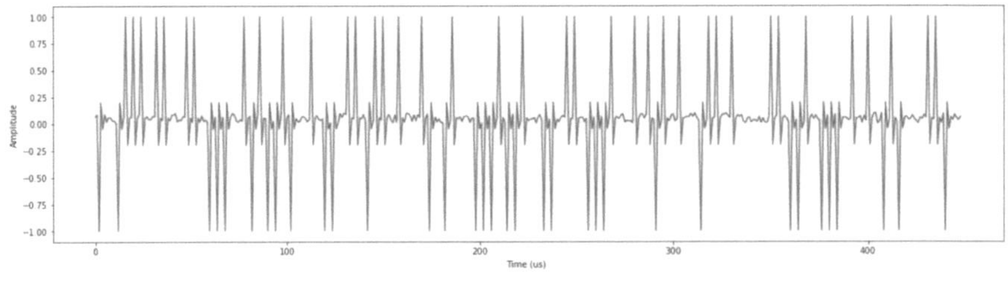

Figure 1.
UWB pulse structure.

it is in most cases as short as possible (to limit the amount of airtime consumed by each ranging frame). In its simplest expression, when the frame is built for ranging between 2 devices, it only contains the header (no payload, no source or destination addresses). In more complex environments (multiple possible senders and receivers), additional fields are added as needed.

The simplest ranging process then consists of an exchange of two frames between 2 ranging-capable devices (RDEVs when implementing 802.15.4a, or enhanced ranging capable-devices - ERDEVs, when implementing improved modes defined in 802.15.4z and described below). The two-frame exchange is simply called Two-Way Ranging (TWR) in 802.15.4a and was renamed Single-Sided Two-Way Ranging (SS-TWR) in 802.15.4z (SS is added in that revision of the Standard to avoid any confusion with another mode, DS-TWR, described in the next session).

One of the devices is called the initiator (in 802.15.4.z, or the originator in 802.15.4a, but the functions are the same in both Standards for the basic TWR case), and the other is the responder.

2.2.2 Basic SS-TWR mode

At time t_0, the initiator (A in **Figure 2**) sends a ranging frame and starts its ranging counter.

The frame travels to the responder (B in **Figure 2**), consuming a time-of-flight t_p that is proportional to the distance between both devices.

Upon receiving the first pulse of the preamble, at time t_1, the responder starts its ranging counter. The responder then receives the rest of the frame, interprets it, and realizes that it is a ranging frame and that it should respond. It then builds a ranging response frame and sends it to the initiator, at time t_2. The responder indicates in the frame a payload value called t_{replyB}, which indicates the time between the reception of the first pulse at the responder antenna (time t_1, at which the responder ranging counter was started) to the time at which the first pulse of the responder's ranging frame is set to leave the responder's antenna (t_2). The responder stops its ranging counter upon sending the response frame.

The frame travels to the initiator, consuming the same time of flight t_p as the first ranging frame (in UWB, one assumption is that devices do not move fast enough from each other for their distance to have significantly changed during the exchange).

The initiator receives the first pulse of the preamble, notes the time from its ranging counter (t_3), then receives the rest of the frame.

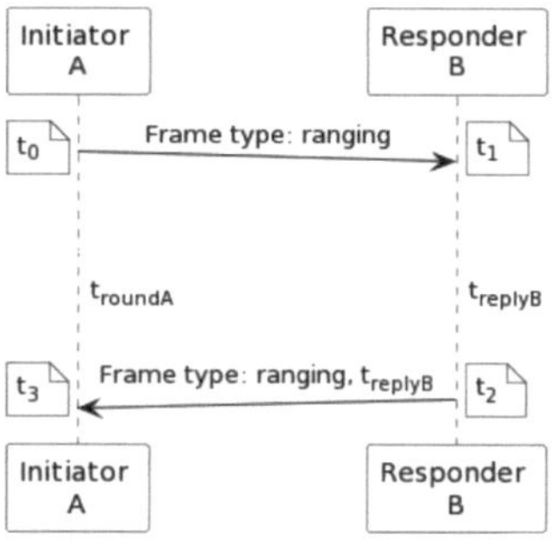

Figure 2.
SS-TWR choreography.

At this point (**Figure 2**), the initiator has the time it took to perform the entire round of exchange ($t_{roundA} = t_3 - t_0$), and the time consumed by the responder's activity (t_{replyB}). As frames traveled both ways, the ToF between devices is simply estimated as:

$$\hat{ToF} = \frac{1}{2}\left(t_{roundA} - t_{replyB}\right). \tag{1}$$

Eq. (1) supposes that both devices' crystals count time at the same speed, as the initiator subtracts t_{replyB} from t_{roundA}. But t_{roundA} is measured by A, while t_{replyB} is measured by B. In most cases, crystals are imperfect and there is a time offset, or drift, between devices. There are multiple proprietary and many common methods to reconcile these differences. For example, one may notice that the PHY header indicates the PRF, and that the duration of each pulse is also known. Therefore, the responder could simply measure each pulse duration and the PRF in the frame received from the initiator, find that they do not perfectly align with its interpretation of the pulse duration and the PRF, and deduce the time offset between its clock and the clock of the initiator.

2.2.3 802.15.4z reply time improvements

802.15.4z allows B to make good use of finding the time offset difference, regardless of how B made that determination. When both A and B support these refinements, they are called ERDEVs.

In an efficient embodiment (SS-TWR with fixed reply time), A sends a ranging frame as in the SS-TWR mode, but A and B agreed in advance (through out-of-band or previous UWB messages) to a specific t_{replyB} value. B receives A's ranging frame and waits for t_{replyB}, then replies. In a sound implementation, B uses the pulse duration and PRF estimation as detailed above to estimate A's clock speed and adjust its t_{replyB} value based on that understanding, therefore attempting to determine t_{replyB} the way A would have calculated it (i.e., "in A's time"). This common understanding limits the effect of the drift.

In another form (SS-TWR with embedded time result, **Figure 3**, left), A, in its ranging request frame, inserts a Ranging Request Measurement and Control Information Element (RRMC IE), that sets a Reply Time Request bit. B then understands that

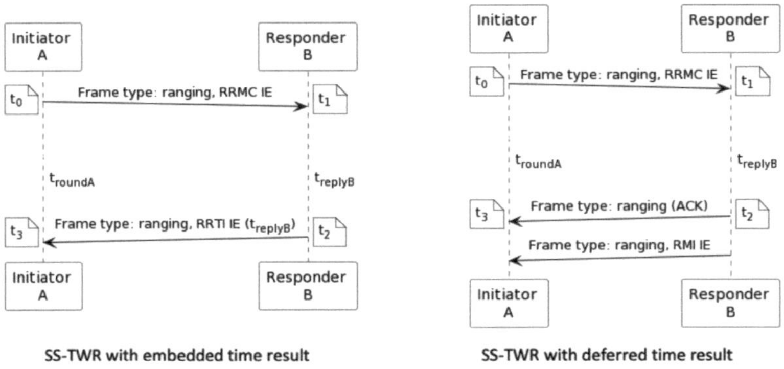

Figure 3.
802.15.4z SS-TWR with embedded time result (left) and 802.15.4z SS-TWR with deferred time result (right).

it must compute the drift directly. Then, in its response frame, B adds a Ranging Reply Time Instantaneous Information Element (RRTI IE), that expresses the calculated time offset. A can then choose to incorporate this drift in its calculation. In the real world, it is unlikely that B would be able to make such drift determination on the fly, but it could compute the offset with better accuracy at each new round. Then, the distance accuracy increases as more rounds are performed between the devices.

In the third and last form (SS-TWR with deferred reply time result, **Figure 3**, right), B would want to provide the offset value in near real-time but does not have the computing capability to calculate its value on the fly. Therefore, upon receiving the ranging request from A, with the Reply Time Request bit set in the RRMC IE, B responds with an acknowledgment frame that allows A to compute t_{roundA}. Then, in a subsequent frame, B sends a Ranging Measurement Information (RMI) Information Element, that includes both t_{replyB} and the estimated offset. The offset between A and B, C_{offs}, is then incorporated into Eq. (1), which becomes:

$$\hat{ToF} = \frac{1}{2}\left(t_{roundA} - t_{replyB} - \left(1 - C_{offs}\right)\right). \quad (2)$$

These improvements allow the ranging exchange to complete with a time error well below 1 ns (often in the 100-picosecond range), allowing a ranging accuracy in the order of 3 cm.

2.3 Double-sided two-way ranging

One limitation of SS-TWR is that responder B does not benefit from the exchange. Its role is merely to respond to A. In the days of 802.15.4a, there was also a concern that clock drifts could not be properly accounted for. The 802.15.4a Standard then devised an additional mode, called Symmetric Double-Sided Two-Way Ranging (SDS-TWR). In this mode, A starts with a ranging frame, as in the SS-TWR basic mode, and B responds with its ranging frame that includes t_{replyB}, but keeps its ranging counter running. Upon receiving the response, A, instead of directly computing the ToF, responds with its ranging frame and processing time (t_{replyA}). B receives that response measures its arrival time and stops the ranging counter. B also measures its own t_{roundB}, which starts when B sends its ranging response frame and stops when it receives A's response. The ToF value is now present 4 times (**Figure 4**): twice in t_{roundA}, measuring the interval from the departure of the first frame from A, and the arrival of the response from B, and twice in t_{roundB}, measuring the interval from the departure of the response frame from B, and the arrival of the response from A.

In this configuration, A can still compute its interpretation of the ToF using Eq. (1). B can also compute its interpretation of the ToF, with the additional advantage, that the error is now reduced as B can compare t_{replyA} to its measurements, and therefore estimate the relative drifts. The estimation remains precisely that, an estimation. However, the process reduces the error. Even with a crystal accurate at 80 ppm, the error in the ToF estimated by B commonly drops well below 10 picoseconds.

802.15.4z added an enhancement to this mode (**Figure 5**), called Double-Sided Two-Way Ranging (DS-TWR, thus without the "symmetric" word in the 802.15.4.a version). In this variation, well adapted to scenarios where the conditions of the channel or other parameters cause t_{reply} to be large on either or both sides (with the consequence that the drift is more difficult to evaluate), A first sends a ranging request, including the RRMC IE and its support for the DS-TWR exchange.

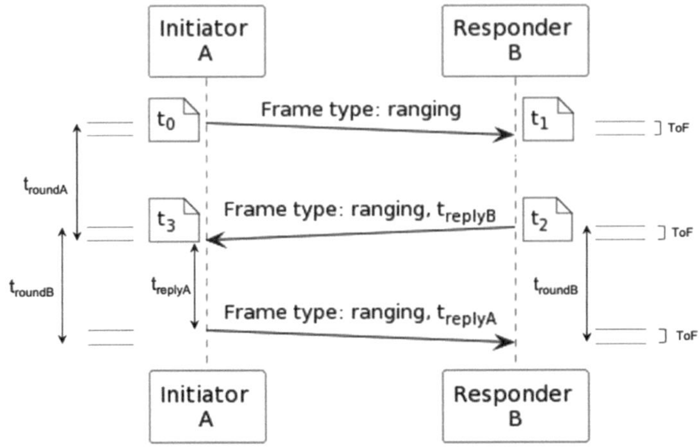

Figure 4.
802.15.4a SDS-TWR.

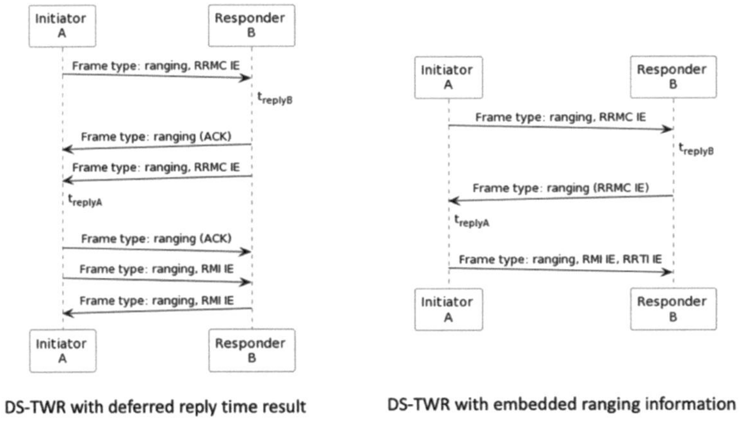

Figure 5.
802.15.4z DS-TWR with deferred reply time result (left), DS-TWR with embedded ranging information (right).

B responds with an Acknowledgement frame, allowing A to measure t_{roundA}. But then, after the round, B initiates its TWR exchange, indicating in the RRMC IE that this is a continuation of the previous exchange. A response with an Acknowledgement frame, allowing B to measure t_{roundB}. Then, in a subsequent frame, A sends another frame to B, that includes the RMI IE and the value for t_{roundA} and t_{replyA}. From all these elements, B can minimize the effects of the drifts and compute the ToF. B can then, in turn, sends a frame to A with the RMI IE that contains t_{roundB} and t_{replyB}. A can then also compute its estimation of the ToF. In both cases, the estimation becomes:

$$\hat{ToF} = \frac{(t_{roundA} t_{roundB}) - (t_{replyA} t_{replyB})}{t_{roundA} + t_{replyA} + t_{roundB} + t_{replyB}} \tag{3}$$

Here again, the error is reduced to less than 10 picoseconds in most cases, and both sides compute the ToF estimate. The main downside of this method is that it requires up to 6 frames to complete. A reduced version of this method is also allowed by

802.15.4z and is called DS-TWR with embedded ranging information. In this variation, B does not respond to A's ranging request frame with an Acknowledgement frame, but directly with a ranging request frame, that includes the RRMC IE to show that the response is a continuation of the exchange. A then responds with a ranging frame that includes the RMI IE and the value of t_{roundA}, but also an RRTI IE that includes the value of t_{replyA}. B can then directly compute its estimation of the ToF. If A also wants to perform this computation, it can set, in the RRMC IE, a field called ToF Request. This causes B, at the end of its ToF estimation, to share the calculated value in a new frame with A (carrying the ToF estimate in the RMI IE).

2.4 Time difference of arrival

All the techniques associated with TWR variants suppose a form of initial configuration between devices, so they know their respective role and place in the sequence of messages. This requirement raises the natural question of the final goal of such a ranging exercise. In many cases, the purpose of the measurement is not to find the relative distance between two objects, but to determine the location of one of them. The use case may then be navigation (a mobile device needs to establish its location, and commonly display it on the local screen over a local map), or asset tracking (a backend management platform needs to record and/or display the location of assets, for example, parts on a factory floor). If the purpose is solely navigation or solely asset tracking, then it becomes practical to deploy a set of devices (now called anchors) at static, known positions, and configure them permanently for navigation or tracking purposes. In this case, the ranging messaging structure can be transformed so that only one side (the mobile device or the static anchors) sends the ranging messages.

This possibility is lightly described in 802.15.4a, and more formally specified in 802.15.4z (under the umbrella name One-Way Ranging [OWR]).

2.4.1 TDoA for navigation

For navigation purposes, the mobile device needs to establish its distance to each known anchor and deduce its location by comparing these distances. In 802.15.4a, the anchors would be statically configured for this purpose. Then, at regular intervals, they would send a message with the Ranging bit set, the Acknowledgement bit not set [so the mobile device knows not to answer], and the identifier of the sending anchor (**Figure 6**). All anchors would send the ranging message at the same time. Because the anchors would be at different distances, the mobile device would receive the messages at different times and would use the time difference of arrival (TDoA) to deduce its location. We will look at this last step in the next section.

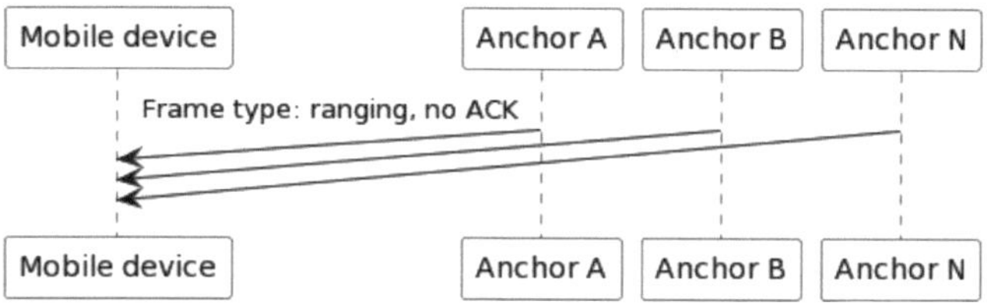

Figure 6.
802.15.4a TDoA Mode 1.

Practically, however, this technique (called TDoA mode 1) made several assumptions that were not always realized:

- The distance between anchors and the mobile device would be different enough that their message would not collide. In the real world, one anchor may be a few meters closer or farther than the other, and the message from one anchor may still be in the process of being received while the message from the other starts arriving at the mobile antenna, causing collisions (resulting in no distance to either anchor).

- Sending the messages at the same time supposes a highly accurate synchronization technique between anchors. 802.15.4a recognized this need but did not propose a mechanism for such synchronization.

802.15.4z describes a variation of this method. In this version, the anchors send their message one after the other, with a precise (and known) offset between transmissions (**Figure** 7). This difference avoids message collisions.

The anchor's clocks still need to be carefully synchronized, and the standard suggests an over-the-wire or over-the-air method, without further details. There are certainly many possible proprietary methods for such exchanges. A practice long established in the industry consists in designating one primary anchor that sends intervals (e.g., every 100 ms) a broadcast message (called a 'sync' message) that includes its time. The others (called the secondary anchors) receive this message and, knowing their distance to the primary anchor (and anchors are static in position and configured by a network administrator), re-align their clock to the primary's time. After a few of these exchanges, the secondary anchors can learn their mean drift over the sync message intervals and re-align their clocks also between sync messages. In stable conditions (e.g., no brutal change of temperature or other operating conditions), the system can reach an accuracy in the 20 to 50 picosecond range. The sync message may also include an ordered list of anchors and an interval. Using that information, the receiving anchors would know which anchor is supposed to send the ranging message first, and how long each next anchor needs to wait before sending its ranging frame. In the real world, as the list of anchors is static, and as one anchor may disappear for any reason, it is common for the network administrator to simply configure each anchor with a time offset ("send your ranging frame $X\mu s$ after detecting anchor i's ranging frame, and/or $Y\mu s$ after detecting anchor j's ranging frame"). Such a static configuration avoids consuming airtime to repeat a sequence

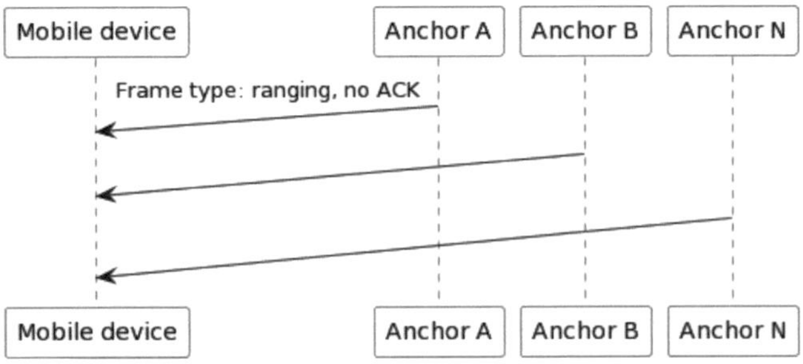

Figure 7.
802.15.4z downlink TDoA.

that is unlikely to change. It also avoids the chain rupture effect of the next anchor never sending a ranging frame, because it is waiting for the previous anchor to send, while that previous anchor was disconnected for some reason.

2.4.2 TDoA for asset tracking

Both 802.15.4a and 802.15.4.z, describe the asset tracking case with similar methods (802.15.4.a calls this case TDoA method 2, 802.15.4.z simply describes it as a second case of TDoA utilization). In this scenario, the mobile device (called the initiator) sends ranging messages (called 'blinks') at regular intervals (**Figure 8**). The header has the Ranging bit set and the Acknowledgement bit not set (no response needed), and the frame can be limited to carrying a form of identifier for the initiator (a MAC address or a simpler identifier), along with a message number. Each anchor individually receives the message, notes the time of arrival, then forwards the message number, initiator identifier, and time of arrival to an external system (usually a Real-Time Location Service [RTLS] server). The server collects such messages from all detecting anchors and uses the time difference of arrival to compute the initiator location.

802.15.4a does not assume the anchors' clocks, but 802.15.4.z recognizes that they must still be synchronized. This is because they need to mention the time of arrival of each blink. If the clocks are not set to the same time reference, these times of arrival cannot be compared. Thus, the asset tracking case still often leverages sync messages sent from a primary anchor to the secondary anchors. A conceptual difficulty associated with this requirement is that the sync messages serve no direct location purpose. They are just part of the necessary mechanic to keep the clocks aligned. Yet they consume airtime, which is not available for the blink messages. Therefore, more frequent sync messages increase the accuracy of the TDoA measurement, but also reduce the possible density of blink messages (thus the number of devices tracked in each space, or the frequency of their blink updates). In most scenarios, an arbitration is made to limit the sync messages to match the inaccuracy tolerance of the location calculation derived from the clock drifts.

2.4.3 TDoA hybrid modes

The binary opposition between asset tracking (the mobile device is the one sending frames used for ranging, the anchors send messages to synchronize their clocks, but are otherwise passively receiving the ranging frames) and navigation (the anchors are the ones sending frames used for ranging, the mobile device is potentially entirely passive and thus invisible to the infrastructure) makes sense in a specialized scenario. However, the real world is often more complex. A department store, for example,

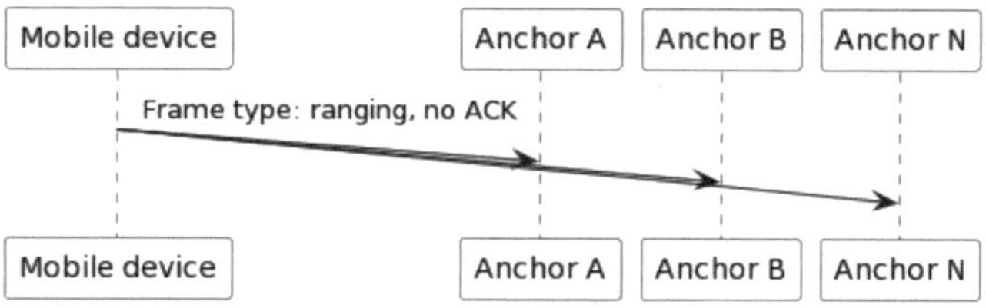

Figure 8.
802.15.4a TDoA Mode 2, 802.15.4z uplink TDoA.

may want to offer ultra-accurate navigation services for its customers, but also track goods and staff in the store. A smartphone might be in the hands of either side (customer or staff). Additionally, the store may want to ensure customer anonymity or may request permission to identify the users using the navigation service. This is where the Standards (802.15.4a and 802.15.4z) stop and where other organizations, like FiRa, define common use cases and specifications among vendors so that the anchors and the mobile device can recognize their operating scenario in the field.

In most cases, the specifications can recognize that UWB is one component of a more complex system, that includes an operating system and possibly other radio technologies (e.g., Wi-Fi or BLE). There is therefore a possibility of signaling (possibly out-of-band, e.g., with BLE or Wi-Fi) between the infrastructure (where the anchors reside) and the mobile device to indicate the scenario:

In the case of anonymous navigation, the infrastructure merely needs to signal the operating parameters (channel and others).

In the case of asset tracking, the infrastructure may provide a form of identifier (this is store X), preferably verifiable by the mobile device (e.g., a hash), and the tracking parameters (interval between frames and others). The mobile device operating system can then parse the message, compare the request to its configuration, and start emitting ranging frames if it is an asset of the store (while a customer mobile device would simply ignore the request).

In a hybrid case, the infrastructure may want to track the device while offering navigation services. Tracking may be a generic analytic need, to observe general movements in the store, without identifying any specific device, or be more specific by tracking individual devices (for example because some customers with a store-specific app may request coupons when in proximity to some type of merchandise). Here again, the infrastructure can signal one or both scenarios. Some mobile devices may then be configured to ignore the request. Others may be configured to only provide anonymous ranging. In that case, the device may (passively) perform navigation, and at random intervals send short series of ranging frames with a temporary (randomized) identifier. As the series is short, sporadic and the identifier randomized, the infrastructure would not obtain more than small snippets of directions, which would not be very useable individually, but would be sufficient, at scale, to provide an understanding of the general movements of people through the store. Other devices may have a specific store app and be configured by the user to provide an accurate location.

There may also be some cases where device tracking becomes mandatory, for example in hazardous areas. Here again, the infrastructure can signal such zones, requesting all devices in the zone to signal their presence. Mobile devices may then be configured to either respond to such requests or ignore them.

2.5 Angle of arrival (AoA)

802.15.4a, published in 2007, did not consider angles. However, several proprietary implementations leveraged the angle of arrival of the signal to deduce its likely direction [2, 3]. This determination has several advantages:

Triangulation or multiangulation (leveraging three, or multiple angles) can complement trilateration, or multilateration (leveraging three, or multiple distances). This point will be explained further in the next section.

When the distance to a single device is evaluated over multiple samples, the observation of the matching angles allows the system to calculate if the signal direction is stable. In an LoS ideal scenario, the source is sending a series of frames (and associated pulses), that all reach the receiver with the same intensity and from the same direction. In an indoor nLoS scenario, some frames may reach the receiver through an LoS path (and their power level and angle can be measured). Others will reflect on obstacles, the LoS signal may be too weak to be detected, and the frames may therefore reach the receiver with different power levels and angles. By comparing the angles of the pulses, from one frame to the next, it is possible to conclude if the channel is stable and LoS, or unstable and/or nLoS. Although leveraging this piece of information in real-time may be difficult, a properly designed system may relate the change of angle with the change in the calculated range value and deduce what is the most likely angle of arrival for the LoS component, if it can be found.

While UWB implementations started leveraging the angle of arrival (under the name Phase Difference of Arrival, PDoA) as soon as 802.15.4.a was published, other radio technologies also started at the same period to consider the angle of arrivals, either at the protocol-definition level (e.g., BLE) or in practical implementations (e.g., Wi-Fi).

This uncoordinated development led to an imprecise terminology that still confuses researchers today. In its most common implementation, UWB (along with BLE or Wi-Fi) considers the angle of a single signal received on two (or more) antennas of a single receiver. The frequency of the signal is known, and consequently its wavelength. For example, UWB channel 9 has its center frequency *f* set to 7987.2 MHz, and its wavelength (*1/f*, usually written λ) is therefore approximately 0.0375 m. By observing the point in the wave cycle (the phase) at which one antenna receives the signal and comparing it to the phase at which another antenna (whose distance to the first antenna is known) receives the same signal, it is then a matter of basic trigonometry to deduce the incident angle of that signal (**Figure 9**).

Several radio technologies call this angle the "signal angle of arrival". Some UWB experts, however, call it "Phase difference of arrival" (pDoA), because it is obtained by comparing the phase between antennas. Unfortunately, several other technologies call pDoA the observation of a signal on a single antenna, comparing the phase of one primary component to the phase of one or more other components (e.g., subcarriers, reflections, etc.) [4]. Other technologies call pDoA the differences in phase of two different signals received on a single antenna [5]. This variability in terminology makes the term pDoA no longer preferred, and AoA may be a safer choice, when in doubt.

Because the consideration of the angle became, in the 2010 decade, an active contributor to many radio technologies seeking location accuracy, 802.15.4z integrates the values. The standard does not define how to measure the angle and merely observes that a system may be able to calculate its value. In all the two-way ranging techniques considered in 802.15.4z, the initiator can request, as an option in the

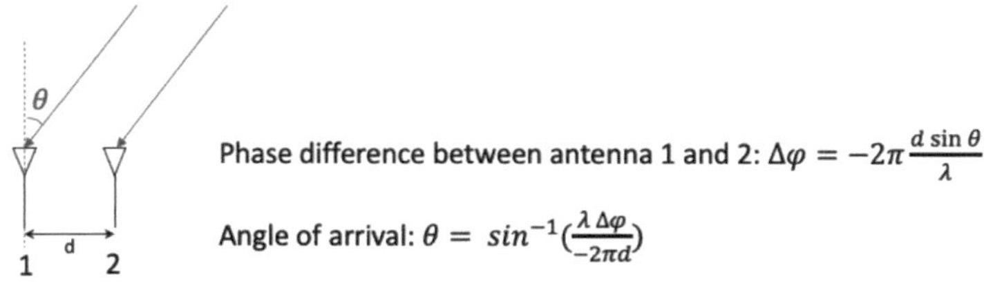

Figure 9.
The angle of arrival determination.

ranging frame, to request for the AoA (azimuth and elevation) at which the frame was received by the responder. The responder indicates these elements (in radians, with a possible range of $[-\pi, \pi]$ for the azimuth, and $[-\pi/2, \pi/2]$ for the elevation) in the RMI IE in the response.

2.6 Protection of UWB exchanges

The UWB frames are not encrypted, and an observer could read the timestamps they carry. When observing the full TWR exchange, the observer may be able to deduce the distance between two ranging UWB devices. The issue is not critical in many settings where the use case is the navigation or asset tracking. However, UWB is also used with TWR for accurate ranging, for example, to automatically open a door when a user is near or unlocking a car. In these cases, an attacker could replay or hijack one side's ranging frames, lead the other side to conclude on a distance shorter than the real physical distance between the initiator and the responder, and thus open the door or unlock the car while the user's device is still far.

802.15.4z designed two mitigation techniques for such an attack. The first form is a mutual authentication between the initiator and the responder, which leads to encrypted exchanges. This mechanism protects from eavesdropping and is well-adapted for scenarios where the initiator and responder know each other (e.g., a car and its key fob). However, an attacker can still hijack the exchanged frames at the PHY level and lure the receiver to conclude on a short ToF. A second mechanism was designed to mitigate this risk, in the form of a Scrambled Timestamp Sequence (STS) field. The STS can be inserted in the physical header of the ranging frames and consists of pseudo-randomized pulses organized in blocks (one to four blocks of 512 chips each $[\sim 1\mu]$, or 128 bits, separated by silences, or 'gaps'). The STS relies on keys that are exchanged in advance between the initiator and the responder, and nonces (numbers used once), from which the transmitter generates a unique value used as the STS to timestamp the ranging frame. The receiver, having the same information, generates the same value, and accepts the ranging frame only if its STS matches the receiver's generated value. Because of the large number of pulse sequences that can be generated for the STS field, the probability that an attacker could generate the right sequence and thus lure the receiver to conclude that the relayed frame did originate from the expected sender, at the expected distance, is minuscule.

Although the STS scheme is not expected to be impossible to attack, it has proven to be robust [6].

3. Finding a location with UWB

Once an estimation of the distance (and possibly angle) between a mobile device and several anchors has been found, the next step is to compare the values and deduce the location of the object. The terminology in this field tends to distinguish the position, which is the conclusion of the comparison of distances (or angles) to known points, from the location, which is the position projected on a known set of references. Thus, for example, one would say that the position of a particular object is at the intersection of three circles of radii 5.7, 8.1, and 7.4 meters from anchors A, B, and C respectively. Then, once the position of the object and the anchors have been projected onto a map, one could declare that the location of the object is on the second floor of the hotel, in the upper left corner of room 242.

Finding the location of the mobile object thus supposes that the location of the anchors is known. For the asset tracking use case, the RTLS server can be configured with such information. In the navigation case, the 802.15.4 Standard does not describe how the mobile object should learn the anchors' location. Conceptually, such information may be embedded in the UWB frame payload. However, the anchors and the mobile device would need to agree on the location format. Industry certifications like the one driven by FiRa define a common format and suggest that the location information should be expressed out-of-band (for example using Bluetooth Low Energy [BLE]).

Regardless of the use case, the first step toward determining location is to compute a position.

3.1 Establishing a position from ranges

3.1.1 Localization in the TWR case

In the TWR cases, ranges are evaluated between a mobile device and a set of anchors. Without angle information, determining that the mobile is x meters away from an anchor places the mobile on a circle of radius x, centered on that anchor. On a plane (i.e., supposing an ideal 2D environment), 2 circles intersect on two points (when they intersect), and 3 anchors are needed to hope for a unique solution (if all three circles intersect). The action of finding the position from three distances is called trilateration. In 3 dimensions, the circles become spheres, and 3 spheres intersect on two points, thus causing position uncertainty. If all anchors are on the same plane, then intersection points are typically above each other. When the position then needs to be projected onto a 2D map, this uncertainty may be acceptable. When 3D representations are needed, and no assumption can be made about the object height, 4 anchors or more are needed to hope to obtain a single intersection point (multilateration).

In an ideal world, the intersection can be simply calculated by solving a set of equations representing the distance of the object to each anchor. If m is a mobile object of unknown coordinates $m = (x_m, y_m, z_m)$, and a_i is an anchor of known coordinates $a_i = (x_i, y_i, z_i)$, the distance between the mobile and the anchor is expressed by the straightforward Euclidean distance equation:

$$d_i^2 = (x_i - x_m)^2 + (y_i - y_m)^2 + (z_i - z_m)^2 \tag{4}$$

Eq. (4) can be re-written in an alternate form:

$$\left(x_m^2 + y_m^2 + z_m^2\right) - 2\left(x_i x_m + y_i y_m + z_i z_m\right) = d_i^2 - \left(x_i^2 + y_i^2 + z_i^2\right) \tag{5}$$

As more anchors (j, k, etc.) are added, it becomes possible to obtain linear equations in x_m, y_m and z_m by subtracting equations pairwise. When 4 such equations are available, the system is overdetermined and a single solution can be found, however, even if UWB allows for high-ranking precision, each measurement is not mathematically perfectly accurate (the experimenter observes an estimation of the range, $\hat{d}_i$, that differs from the true range by an unknown factor ϵ_i so that $\hat{d}_i = d_i + \epsilon_i$) and in the real world, a perfect solution cannot be found.

3.1.2 Localization in the TDoA case

The same type of issue can be observed in the TDoA case. With OWR, the detecting side never measures distances directly. Instead, the observation is mere that a given mobile signal arrives on an anchor *n us* earlier than on another anchor (or equivalently that the mobile received the signal from one anchor *n us* earlier than the signal from another anchor). As the speed of light and RF signals are known, this observation is translated into the conclusion that one anchor is *x cm* closer than the other one, but the distance itself is not known. This relationship translates into a hyperbolic line between anchors (**Figure 10**). TDoA measurements between two anchors form one hyperbola.

Conceptually, three anchors result in 3 hyperbolae. However, in the real world, the TDoAs are compared against a primary anchor (e.g., anchor 1) and the hyperbola that compares anchor 2 to anchor 3 is unusable. Therefore, three anchors result in 2 usable hyperbolae, which intersect at a single point on a 2D plane. For 3D determination, at least 4 anchors are therefore needed. The comparison is then translated into a matrix of compared distances, from which the true Euclidean distance can be found iteratively [4]. Here again, distances in the real world are noisy, and no perfect solution can be found. TDoA also suffers from the additional difficulty that, contrary to circles, hyperbolae are asymptotic to a line. Practically, this fact translates into the issue that, as the observed compared distance deviates from the ground truth, it worsens faster (up to infinity) than its TWR counterpart.

3.2 Least square solutions

To solve the issue of reconciling noisy distances, a natural approach is to attempt to determine the measurement errors and minimize them. Mathematically, with N anchors, this requirement is expressed as

$$\min_{m} \sum_{i=1}^{N} \left(||m - a_i|| - \hat{d}_i \right)^2 \tag{6}$$

Naturally, the experimenter does not know m but can find the coordinates iteratively, with techniques like gradient descent where the derivative of Eq. (6) for each

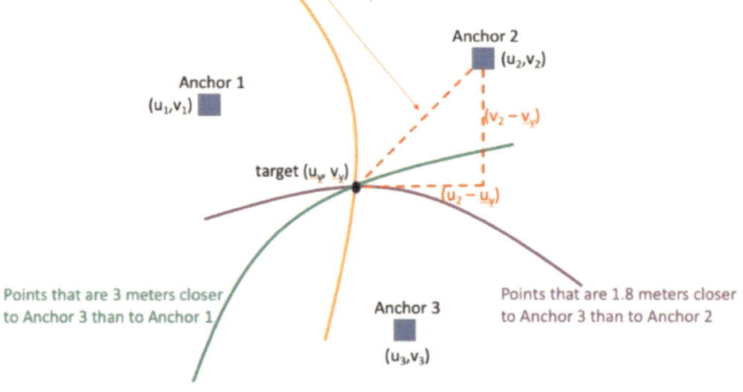

Figure 10.
Hyperbolae formed from OWR measurements.

component of $m(x_m, y_m, z_m)$ is calculated, then an iterative process tests the values of each component until the m that minimizes that derivative is found.

This least square (LS) method has been well established for noisy distance resolution when the errors to all anchors are comparable. In some environments, however, some anchors are in direct LoS to the measuring mobile device, while some others are behind obstacles. In this type of scenario, fusion techniques appear, where LS is complemented with steps that either estimate the nLoS deviation [7] to allow anchor-to-anchor comparison or use fingerprinting techniques [8] to place the mobile at the position of best likelihood.

3.3 Bayesian framework solutions

One key aspect of the LS approaches is that they examine each measurement set individually. However, it may be tempting to reason that a mobile device is by nature moving, and that the position at time t_{n+1} may be related to the position at time t_n. Therefore, localization determination often borrows from Bayesian filtering techniques, where the state of a dynamic system is evaluated from noisy measurements and compared to the conclusion of previous measurements. There are naturally many techniques falling into that family, and this chapter only underlines the most used.

3.3.1 Kalman filter

Among the techniques leveraging the Bayesian framework, the most common is the Kalman filter (KF). The technique has been used successfully since the 1960s for trajectory and position estimation. In essence, the standard KF approach compares the estimation at time t_n with the prediction built from past observations and predictions [9]. When the new observation is noisy, the algorithm trusts more (affects a higher weight to) the prediction. When the new observation noise is small, the algorithm affects a higher weight to that observation than to the prediction.

One key requirement of the KF is that the underlying equations must be linear. The noise statistic also must have a Gaussian distribution. For location estimation, the distance equations are commonly not linear, and researchers have proposed several extensions to the KF for these cases. The most common variants are the unscented Kalman filter (UKF) and the extended Kalman filter (EKF). Both models solve the non-linearity by approximating the nonlinearity with Taylor expansions, then estimating their derivatives (which then become linear equations), EKF with the first derivative, and UKF with the second derivative. EKF is commonly used when the noise figure is small (variance differences are not large, mostly in LoS environments). UKF appears often in scenarios where the noise is high (environments dominated by nLoS scenarios).

The KF family has received the favor of implementers of indoor localization algorithms because, despite its complexity, the KF relies on matrix operations that most operating systems integrate natively. Thus, the computation can be done efficiently on most systems, in near real-time, and KF techniques are widely successful for UWB-based localization [10–12]. However, you should be aware of several limitations:

The KF methods require an initial position estimation, which in most cases is not available (and thus a random or arbitrary value is fed into the system). The algorithm then converges as more observations are made. The pace of these observations has a direct influence on the convergence speed. In other words, if your UWB method observes one position per second, the convergence to an accurate-enough position will be much slower than if your system generates 50 measurements per second. A common implementation practice is therefore to ignore the first *n* estimations to give time for the system to converge.

The KF methods are sensitive to sudden changes. By their very nature, they affect a low weight for measurements that are very far from the estimated positions. But these positions are based on past observations. A direct consequence is that the trajectory is linear, and the mobile device suddenly changes direction (for example, the user turns a corner in a corridor), the KF methods tend to overshoot, estimating that the new observation is likely inaccurate. On a map, you would then see the device trajectory continuing for a little while (possibly through a wall) before slowly turning and catching up to the user's real position. Here again, the pace of the sampling dictates the duration and span of this negative effect.

3.3.2 Particle filters

Particle filter (PF) is the name given to a family of techniques that implement the Monte Carlo approach within the Bayesian framework [13]. It is well adapted for non-Gaussian, non-linear estimations, and thus often used for indoor ranging problems. This is because, in LoS conditions, the observations may display some noise value forming a Gaussian distribution around some value, but in nLoS conditions, walls and multipath tend to inflate the observed values, causing a long tail of distance overestimations that negates the Gaussian condition.

At its core, the PF incorporates two models. A motion model reads a set of values and deduces a possible state. In the case of indoor location, this first set may be obtained for example from odometry (readings from the device's internal sensors) to estimate the device's new position. In most cases, this technique alone is not sufficient to compute the path, because the sensors operate at a small scale and their accuracy suffers at a larger scale. For example, suppose a smartphone gyroscope estimates an 81-degree left turn, but the user turned 92 degrees left. After walking 10 more meters, the system sees the device 2 meter right of its real position. As the user moves, the gyroscope, accelerometer, and all the other sensors get multiple and frequent inputs, and their errors tend to build on one another, making that odometry can be quite accurate at a small scale (it is called a local technique), for small movements, but is not a great tool to compute trajectory at a large scale if the real position of the mobile is not rectified at intervals, using another source of truth (called a global technique), that may not be very accurate at small scale but may provide better visibility at these larger scales. Outdoor, that source can be GPS. Indoor, a ranging method is a great second source.

PF, therefore, includes, in addition to the motion model, a sensor model that measures the distance to some reference points, with the goal of re-positioning the mobile device from this second source. Naturally, the second source alone is not sufficient (otherwise there would be no need for the first model), as measurements are also noisy. So, PF operates by collecting a set of observations (in our case, distances to anchors), that are called particles, and establishing their probability density function (PDF) to determine which of the observations are most likely to be correct. PF is not very well-adapted to high-dimension problems, but works well for indoor localization, and is therefore widely used with UWB TWR [14, 15], either alone, or in combination (or in comparison) with KF [16, 17].

3.3.3 Machine learning methods

The methods examined so far rely on the laws of physics to find the best range estimation for each anchor and deduce the best position from the combination of ranges available. With multiple samples from multiple anchors, the number of parameters becomes large enough that statistical methods can be successfully substituted for physical methods. These complementary approaches help address two types of issues:

nLoS detection: LoS measurements are closer to the ground truth distance than nLoS measurements. Detecting nLoS conditions (and the stretch they induce to the measured distance) has been an active field of research, where unsupervised techniques can dramatically help group alike nLoS scenarios [18, 19] and reduce the effect of the stretch they induce.

Insufficient contributions: when there are not enough anchors to range against, or when they are all nLoS, supervised techniques allow the operator to sample measurements in different known locations, then deduce the location matching a new set of measurements by comparing it to the sample values. This technique is commonly called fingerprinting and is used on its own [20], or in combination with other techniques [21].

4. Conclusion

This chapter examined the evolution of UWB Standards for ranging. First defined in the IEEE 802.15.4a amendment in 2007, at a time when other groups also claimed ultra-wideband transmissions, UWB initially focused on the simple case of TWR, where one initiator would range against one responder. The integration into a larger localization solution implied a static configuration of roles and relied on implementers to fill the elements undefined in the Standard.

As UWB proved an efficient technology for accurate ranging, multiple proprietary implementations appeared that leveraged the basic tools defined in the protocol, but also added improvements and new modes to better fulfill the different location use cases. In 2020, IEEE 802.15.4z integrated many of these elements, to better address the challenges of TWR implementation, and account for the most popular augmentations, namely TDoA and AoA.

The Standard defines elements of the Physical and the Data Link layers and is silent on how UWB should be used in an end-to-end localization solution. It is therefore not sufficient for practical implementation. Organizations like FiRa integrate the IEEE protocol into a larger landscape, addressing the various use cases and the required communication structure above the two bottom layers.

This combination has made UWB very successful for localization with ultra-high accuracy. The techniques of converting a series of ranges, angles, or time of arrival differences into a position are not specific to UWB. However, the precision allowed by the structure of the UWB signal makes it a prime candidate to solve complex indoor navigation and asset localization problems, especially in the world of robotics and the Internet of Things (IoT).

The journey is far from over. IEEE 802.15.4z assumes that many elements required by the ranging exchange are sent out-of-band or consume in-band airtime that could be used to perform more or better ranging. There is therefore still a need to refine the technique and integrate it with the standard more tools to simplify the communication or perfect the accuracy of the range obtained. The IEEE 802.15ab task group has been formed to tackle this task, with the ambition to complete its work by the end of 2025.

Author details

Jerome Henry
Cisco Systems, Research Triangle Park, USA

*Address all correspondence to: jhenry@ieee.org

References

[1] IEEE. 802.15.4a. Part 15.4: Wireless Medium Access Control (MAC) and Physical Layer (PHY) Specifications for Low-Rate Wireless Personal Area Networks (WPANs), Amendment 1: Add Alternate PHYs. New York. 2007

[2] IEEE. 802.15.4z. Part 15.4: Wireless Medium Access Control (MAC) and Physical Layer (PHY) Specifications for Low-Rate Wireless Personal Area Networks (WPANs), Amendment 1: Enhanced Ultra Wideband (UWB) Physical Layers (PHYs) and Associated Ranging Techniques. New York. 2020

[3] Malajner M, Planinšič P, Gleich D. UWB ranging accuracy. 2015 International Conference on Systems, Signals and Image Processing (IWSSIP). London. UK. 10-12 Sept. 2015. pp. 61-64. DOI: 10.1109/IWSSIP.2015.7314177

[4] Sackenreuter B, Hadaschik N, Faßbinder M, Mutschler C. Low-complexity PDoA-based localization. 2016 International Conference on Indoor Positioning and Indoor Navigation (IPIN). Madrid. Spain. 4-7 Oct. 2016. pp. 1-6. DOI: 10.1109/IPIN.2016.7743692

[5] Naz A, Asif HM, Umer T, Kim B-S. PDOA based indoor positioning using visible light communication. IEEE Access. **6**:7557-7564

[6] UWB Secure Ranging in FiRa [Internet]. Available from: https://www.firaconsortium.org/sites/default/files/2022-09/FIRA-Whitepaper-UWB-Secure-Ranging-August-2022_0.pdf

[7] Yu K, Wen K, Li Y, Zhang S, Zhang K. A novel NLOS mitigation algorithm for UWB localization in harsh indoor environments. IEEE Transactions on Vehicular Technology. 2019;**68**(1): 686-699

[8] Poulose A, Eyobu OS, Kim M, Han DS. Localization error analysis of indoor positioning system based on UWB measurements. 2019 Eleventh International Conference on Ubiquitous and Future Networks (ICUFN). Zagreb. Croatia. 2-5 Jul. 2019. pp. 84-88. DOI: 10.1109/ICUFN.2019.8806041

[9] Welch GF. Kalman filter. In: Computer Vision. Cham (SW): Springer; 2020. DOI: 10.1007/978-3-030-03243-2_716-1

[10] Fu J, Fu Y, Xu D. Application of an adaptive UKF in UWB indoor positioning. 2019 Chinese Automation Congress (CAC). Hangzhou. China. 22-24 Nov. 2019. pp. 544-549. DOI: 10.1109/CAC48633.2019.8996692

[11] Feng D, Wang C, He C, Zhuang Y, Xia X. Kalman-filter-based integration of IMU and UWB for high-accuracy indoor positioning and navigation. IEEE Internet of Things Journal. 2020;7(4): 3133-3146

[12] Cano J, Chidami S, Ny J. A Kalman filter-based algorithm for simultaneous time synchronization and localization in UWB networks. 2019 International Conference on Robotics and Automation (ICRA). Montreal. Canada. 20-24 May 2019. pp. 1431-1437. DOI: 10.1109/ICRA.2019.8794180

[13] Elfring J, Torta E, van de Molengraft R. Particle filters: A hands-on tutorial. Sensors. 2021;**21**(2):438

[14] Yang W, Zhang W, Li F, Shi Y, Nie F, Huang Q. UAPF: A UWB aided particle filter localization for scenarios with few features. Sensors. 2020;**20**(23):6814

[15] Li Z, Wu J, Kuang Z, Zhang Z, Zhang S, Dong L, et al. Moving target

tracking algorithm based on improved resampling particle filter in UWB environment. Wireless Communications and Mobile Computing. 2022;**2022**: 9974049

[16] Petukhov N, Zamolodchikov V, Zakharova E, Shamina A. Synthesis and comparative analysis of characteristics of complex Kalman filter and particle filter in two-dimensional local navigation system. 2019 Ural Symposium on Biomedical Engineering, Radioelectronics and Information Technology (USBEREIT). Yekaterinburg. Russia. 25-16 Apr. 2019. pp. 225-228. DOI: 10.1109/ USBEREIT.2019.8736595

[17] Li X, Wang Y, Liu D. Research on extended Kalman filter and particle filter combinational algorithm in UWB and foot-mounted IMU fusion positioning. Mobile Information Systems. 2018;**2018**: 1587253

[18] Krishnan S, Xenia Mendoza Santos R, Ranier Yap E, Thu Zin M. Improving UWB based indoor positioning in industrial environments through machine learning. 2018 15th International Conference on Control, Automation, Robotics and Vision (ICARCV). Singapore. 18-21 Nov. 2018. pp. 1484-1488. DOI: 10.1109/ ICARCV.2018.8581305

[19] Fan J, Awan A. Non-line-of-sight identification based on unsupervised machine learning in ultra wideband systems. IEEE Access. 2019;7: 32464-32471

[20] Che F, Ahmed A, Ahmed QZ, Zaidi S, Shakir M. Machine learning based approach for indoor localization using ultra-wide bandwidth (UWB) system for industrial internet of things (IIoT). 2020 International Conference on UK-China Emerging Technologies (UCET). Glasgow. UK. 20-21 Aug. 2020. pp. 1-4. DOI: 10.1109/UCET51115.2020. 9205352

[21] Poulose A, Dong S. UWB indoor localization using deep learning LSTM networks. Applied Sciences. 2020; **2020**(10):62-90

Chapter 5

Toward UWB Impulse Radio Sensing: Fundamentals, Potentials, and Challenges

Jonas Ninnemann, Paul Schwarzbach and Oliver Michler

Abstract

Radio sensing is a rapidly emerging research field. It focuses on designing an integrated communication system that can also perform localization and radar functionalities sharing the same transmit signals and potentially the same hardware. Ultra-wideband (UWB) impulse radio is a promising technology for radio sensing because it offers a high-range resolution and direct access to the channel impulse response (CIR) to observe the multipath components (MPCs) of the wideband channel caused by scattering at target objects. This approach enables a wide range of functionalities and applications, especially in the field of mobility and transportation. The foundation is given by the signal propagation and channel modeling of the UWB channel, which is briefly revisited in this chapter. Based on the CIR and estimated MPCs the target object can be localized like a multistatic passive radar. The influence of geometry in a passive target localization system is studied by calculating the geometric dilution of precision (GDOP). In addition to passive localization more tasks and functionalities of radio sensing, are briefly introduced including detection, tracking, imaging, counting, and classification. The chapter concludes with further research directions and challenges in UWB radio sensing, especially for real-world use in the context of mobility applications.

Keywords: multipath-assisted radio sensing (MARS), channel model, channel impulse response (CIR), multipath components (MPCs), impulse radio (IR), radio sensing for intelligent transportation systems (ITS)

1. Introduction

Ultra-wideband (UWB) impulse radio (IR) is extensively researched and used as a technology for indoor positioning systems [1, 2] and short-range communication [3]. Such systems are enabled by key features of the UWB physical layer (PHY) such as high bandwidth, the transmission of very short impulses, and low power. These features lead to a high range resolution and usually dense networks, which in general makes UWB perfectly suitable for radio sensing tasks and applications.

The term radio sensing first emerged in the context of cellular networks and is referring to the usage of existing radio signals to passively sense the environment [4–6].

The goal is to perform communication, localization, and radar functionalities by sharing the same transmit signals and potentially the same hardware. The analysis of the radio signal itself enables new features and functions of the communication systems such as localization, tracking, imaging, detection, or classification of passive target objects without the need for dedicated hardware or specialized measurement setups. In the past, this approach referred to UWB-IR and was mostly considered for tracking [7, 8], imaging [9], and people counting [10]. In contrast to such radar systems, radio sensing follows a more integrated approach by combining communication, localization, and sensing functionalities with the same radio signal and hardware to be more cost and spectrum efficient. This way the IR could be an integrated part of future communication networks to fulfill sensing tasks [5].

UWB and its IR nature allow direct access to the channel impulse response (CIR) in the time domain as the fundamental signal parameter for channel estimation. The CIR measurement includes information about the different propagation paths of the signal (direct path and echo paths). For passive target localization, the CIR is measured, and the multipath components (MPCs) are extracted to fulfill the different tasks of radio sensing. The high bandwidth of UWB ($\geq$ 500 MHz) allows the distinguishing of MPCs in the CIR, even if the time delay of the propagation paths is relatively close together [11].

There are many use cases of radio sensing ranging from smart cities, smart homes, vehicular networks, and health to drones. In terms of mobility applications, radio sensing based on UWB could be helpful and game-changing for intelligent transportation systems (ITS). **Table 1** lists possible use cases of radio sensing for the various modes of transport and tasks of radio sensing.

To implement the use cases, UWB is a promising technology for radio sensing. Therefore, this chapter transfers radio sensing approaches to UWB for different tasks and functionalities. In particular, the chapter provides the necessary fundamentals and proposes a UWB radio sensing approach based on the multipath channel model. Limitations in terms of range resolution and sensor arrangement are discussed, as well as further research directions and challenges.

The rest of this chapter is organized as follows: Section 2 gives an overview of the research field of radio sensing, the different scientific and technological influences, and the current state-of-the-art. In Section 3, the basics of signal propagation are introduced to derive the wideband multipath channel model. Based on this, a UWB radio sensing approach is then described in Section 4 for passive target localization. In addition, the influence of network geometry is investigated for different transceiver

Mode of transportation	**Localization**	**Detection**	**Classification**
Automotive	Vehicular networks and automotive radar	Parking lot occupancy detection	Automotive radar
Rail transport	Flow management in stations	Passenger detection on platforms or inside trains	
Aviation	Boarding monitoring	Seat occupancy detection	Classification of objects to automated passenger safety checks
Drones/UAVs	Landing system	Obstacle detection	Obstacle classification

Table 1.
Use cases of UWB radio sensing for intelligent transportation systems.

constellations and other tasks of radio sensing are briefly introduced. In Section 5 challenges and further research directions for UWB-based radio, sensing is discussed. The chapter concludes with a summary of the key contributions in Section 6.

2. Taxonomy of radio sensing

The research field of radio sensing is composed of different scientific directions and is only made possible by the fusion of these influences and ideas. The three main pillars are wireless communication systems, localization based on RF signals, and radar systems. The different terms and concepts in this highly interconnected research field are presented in a word cloud in **Figure 1**.

Wireless communication systems [12, 13] have been used not only for networking but also for the localization of devices for years now. Especially in the context of indoor positioning systems (IPS) [1, 2], where no global navigation satellite systems (GNSS) are available, localization enables new services and applications. To estimate the location of a mobile sensor (tag) a wireless sensor network (WSN) [1], consisting of fixed sensors (anchors) with known positions, is placed in the environment. Based on different channel parameters and positioning principles, the position of the mobile tag is estimated. Because the target object or user needs to wear an active sensor, this technique is referred to as active localization. UWB is one technology for such localization systems, perfectly suitable due to its high positioning accuracy.

The same radio signal used for active localization can also be analyzed to enable device-free passive localization (DFPL) [14, 15]. Here, the target object does not need to carry a sensor, but instead the position is estimated by evaluating different channel and propagation effects. DFPL is one possible task of a radio sensing system. But based on the use of wireless radio technology and channel estimation procedures, many tasks can potentially be accomplished. For example, detection of the target object, mapping of the environment, tracking, classification of different scenes or counting of objects and people.

In the past, such tasks were accomplished by dedicated radar systems with specific hardware, spectrum, and techniques for the measurement of radar parameters. In

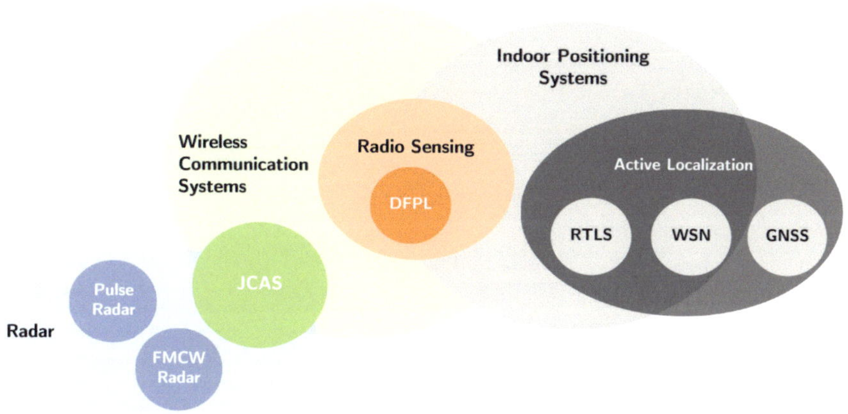

Figure 1.
Taxonomy of radio sensing as a fusion of different research directions in the context of integrated wireless communication systems. New abbreviations are joint communication and sensing (JCAS), frequency-modulated continuous wave (FMCW) radar, and real-time locating systems (RTLS).

principle, there are two different types of radars: continuous wave (CW) and pulse radar. One specific implementation of a CW radar is the frequency-modulated continuous wave (FMCW) radar. There are different geometrical configurations of radar systems such as monostatic, bistatic, or multistatic systems. The radar system can detect objects by transmitting a pulse or CW signal toward the target object and analyzing the reflected signal to estimate different signal parameters like the time-of-flight to the object [16].

More recent research focuses on the integration of these dedicated radar systems into cellular communication systems such as 5G/6G. The integration can be achieved through different levels starting from a better spectral coexistence of radar and communication systems over uniform hardware, RF frontend, and waveform design to true perceptive networks in the future. Different terms are used to describe such systems in research like Joint Communication and Sensing (JCAS) [5], Integrated Sensing and Communication (ISAC) [17], or radar communication (RadCom) [18]. This integration is only possible because of the trend to higher frequency spectrum and bandwidth in cellular networks and the resulting higher range resolution for sensing applications.

Paper	Channel parameter	Approach and algorithm	Task	Evaluation and use case
[19]	Amplitude and phase of the CIR	Multipath-assisted device-free localization	Mapping and localization	Simulation and measurements with DW1000[a]
[11]	CIR	Multipath-assisted radio tomographic imaging	Mapping and localization	Measurements with DW1000
[20]	MPCs	Multipath-enhanced device-free localization and Bayesian localization	Localization	Measurements with DW1000[a]
[21]	MPCs	Channel-SLAM with virtual transmitters	SLAM	Simulation
[7]	CIR	Background subtraction and particle filter	Real-time tracking	Measurements with DW1000[a]
[22]	Channel frequency response	Classification	Human activity recognition	Measurements with DW1000[a]
[23]	MPCs	Multipath-assisted positioning	Localization	Computational results
[24]	MPCs	Multipath-assisted navigation and tracking with PHD-filter	Multiple target tracking	Indoor UWB measurements
[25]	CIR and MPCs	Convolutional neural networks, particle filter	Passive human tracking	Measurements with DWM1000[a]
[26]	CIR	Clustering and Maximum Likelihood Equation	People counting	Measurements

[a]*DW1000 and DWM1000 is a UWB Transceiver IC and module from Qorvo [27].*

Table 2.
Different approaches and measurement principles for UWB radio sensing.

In terms of UWB, the IR is already part of the PHY concept. UWB sends very short pulses over a large bandwidth and uses pulse position modulation to transfer information. This concept enables active localization based on time of arrival estimation. Based on the transmitted pulses and the impulse response of the wideband channel, the PHY of UWB is perfectly suitable for different sensing tasks. In research, different approaches, algorithms, and use cases for UWB sensing are currently discussed (**Table 2**).

3. Signal propagation and channel model

3.1 Propagation phenomena

A radio communication system emits an electromagnetic wave, that experiences different propagation phenomena and effects before reaching the receiver (RX). In general, the energy of the radio signal is reduced in free space, where no obstacle is located between the transmitter (TX) and the RX (**Figure 2**). The Friis transmission model [28] stated that the received power P_r is given as follows [29]:

$$P_r = P_t G_t G_r \left(\frac{\lambda}{4\pi d} \right)^2 \tag{1}$$

where P_t is the transmitted power, G_t and G_r are the antenna gain of the TX and RX. The free-space path loss (FSPL) depends on the wavelength λ and traveled distance d of the signal.

In addition to the FSPL, the received power is further reduced, and the signal direction is influenced by three other basic propagation phenomena: reflection, diffraction, and scattering (**Figure 2**). These effects occur when the electromagnetic wave with wavelength λ encounters an object with size A. Reflections appear when the size of the object is very large compared to the wavelength: $A \gg \lambda$. The angle of the

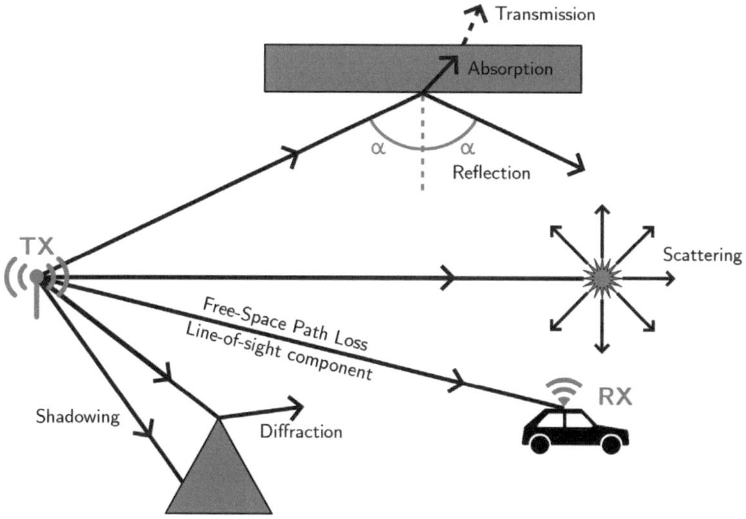

Figure 2.
Signal propagation and phenomena.

incident wave concerning the surface normal is equal to the angle that the reflected wave makes to the same normal. Reflections lead to a decrease in received power due to absorption or even the transmission loss of the wave energy by the encountered object. Diffraction arises when the wave hits an object with a size in the order of the wavelength ($A \approx \lambda$) and is explained by the Huygens principle. Scattering is the result of the encounter with a very small object compared to the wavelength: $A \ll \lambda$. Scattering occurs on objects with rough surfaces, whereby the incident wave is redirected in many directions [29, 30].

3.2 Channel effects

The wireless channel is affected and characterized by the variation of the channel strength or energy level over time and frequency. All effects combined are called fading, and the summarized attenuation of all effects is the path loss between the TX and the RX. Fading can mainly be categorized into large-scale fading and small-scale fading, according to **Figure 3**. Large-scale fading characterizes the variations in path loss over distance (FSPL, log-normal), and also shadowing (slow fading) or even blockage by large objects. These effects are typically frequency independent. Small-scale fading on the other hand describes the constructive or destructive interference of multiple signal paths between the TX and RX. This is caused by scattering and generally leads to a time-varying channel [13, 31, 32].

Especially small-scale fading must be considered for wideband channel models targeting sensing applications. Small-scale fading results in either a flat fading channel or a frequency-selective channel. The coherence bandwidth B_c is considered the bandwidth where the channel is regarded as a flat channel. This mean all signals passing through the channel experience similar attenuation and phase shifts. The root mean square (RMS) delay spread τ_{rms} is inversely proportional to B_c, which means larger τ_{rms} results in a more frequency selective fading channel. If the signal bandwidth B_s is higher than B_c the channel is considered a frequency selective fading channel. The time interval over which the wireless channel is constant is called coherence time T_c [29, 30].

3.3 Multipath propagation channel model

The path loss is the basis for empirical channel models, which model the received power at a reference distance according to the carrier frequency and the environment [29]. Empirical models consider different types of propagation environments and are

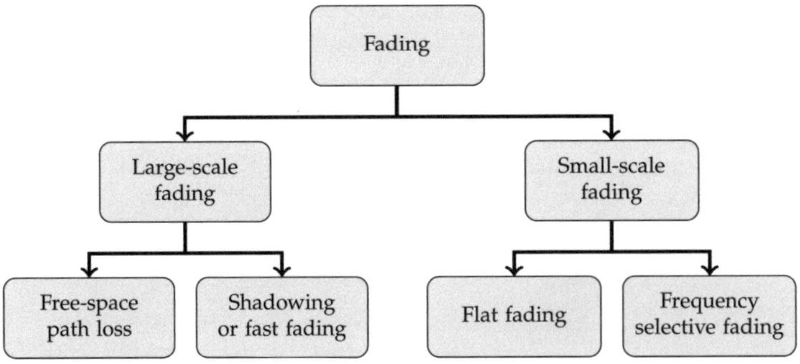

Figure 3.
Types of channel fading effects.

based on real-world measurements [33]. Flat fading channels are modeled with statistical models. For example, the Rayleigh fading distribution describes a statistical time-varying model for the propagation of electromagnetic waves. The Rice distribution modeling, a channel with one strong LOS component [29]. These models are based on measurements of the channel statistics for different predefined environment categories, but fail to resolve individual propagation paths. In comparison to statistical and empirical models, a deterministic model is more suitable for sensing approaches and applications. Individual propagation paths are calculated based on the aforementioned channel effects.

A multipath wideband frequency selective channel can be modeled as a linear time-varying system. We assume that the attenuation and propagation delay does not depend on the frequency inside the range of the coherence bandwidth of the channel. We can generalize the system to an arbitrary input $x(t)$ and compute the received signal $y(t)$ as follows:

$$y(t) = \sum_{i=1}^{N} a_i(t)x(t - \tau_i(t)) \tag{2}$$

where N is the number of different propagation paths with an attenuation $a_i(t)$ and propagation delay $\tau_i(t)$ at time t [13].

Because the channel is linear, it can be described by an impulse response $h(\tau, t)$ [13]. The CIR of a time-varying multipath channel is given as [29]:

$$h(\tau, t) = \sum_{i=1}^{N} a_i(t)\delta(t - \tau_i(t)) \tag{3}$$

where $a_i(t)$ and $\delta(t - \tau_i(t))$ is the attenuation, respectively, the delta function of the ith delayed propagation path. The Fourier-transformed impulse response of the system results in the following frequency response $H(f;t)$ in frequency domain f [13]:

$$H(f;t) := \int h(\tau, t)e^{-j2\pi f\tau}d\tau = \sum_{i=1}^{N} a_i(t)e^{-j2\pi f\tau_i(t)} \tag{4}$$

The fading multipath channel is now described by an input/output relation as an impulse response of a linear time-varying system. The system can be interpreted as a linear finite impulse response (FIR) filter and is also referred to as the tapped delay line model. An example of such an FIR-based channel model is illustrated in **Figure 4** with three different reflected paths and the LOS path, resulting in a four-tap FIR filter. Each tap corresponds to a reflected path with an amplitude $a_i(t)$ and a corresponding delay $\tau_i(t)$.

In a stationary case, where the $a_i(t)$ and $\tau_i(t)$ do not depend on time t we can model the channel as a usual linear time-invariant (LTI) system with the CIR corresponding to the following equation:

$$h(\tau) = \sum_{i=1}^{N} a_i\delta(t - \tau_i) \tag{5}$$

Multipath propagation causes different propagation effects depending on the propagation paths and shadowing. The direct path between the signal TX and RX is referred to as line-of-sight (LOS) propagation. In contrast, an obstructed or reflected

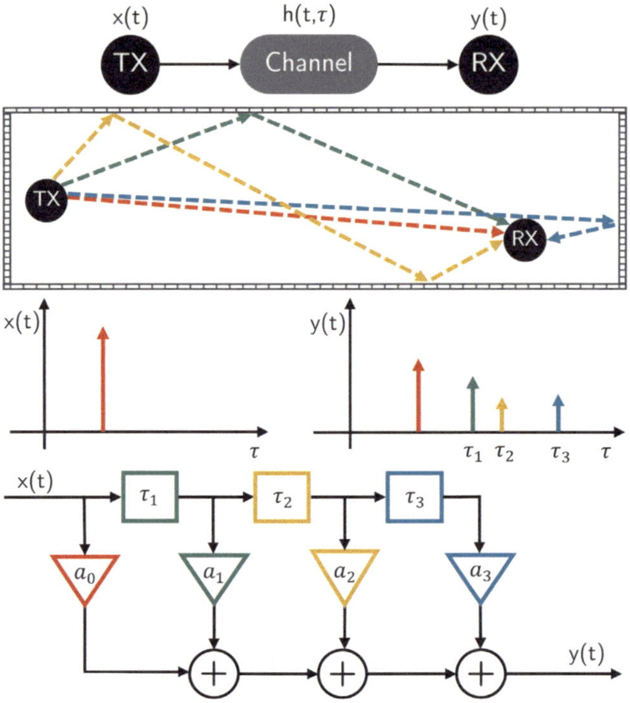

Figure 4.
Tapped-delay-line representation of the time-variant multipath channel model.

transmission path is called non-line-of-sight (NLOS). The term multipath reception applies if the signal reaches the RX via multiple paths caused by different propagation phenomena. This results in a received signal composed of attenuated, delayed, and phase-shifted replicas of the transmitted signal. These components can take different paths in the environment before reaching the RX and are thus called multipath components (MPCs) [30, 32].

3.4 Bandwidth and range resolution

The next step in channel modeling is to convert the time-continuous channel to a time-discrete channel with limited bandwidth B_s. In the case of UWB, the input waveform of the channel or transmitted signal is a Gaussian pulse with a certain pulse duration $T_d = \frac{1}{B_s}$ and a signal bandwidth of $B_s \geq 500$ MHz. The rectangular shape in the frequency domain corresponds in the time domain to the sinc function. Based on the sampling theorem we can sample the CIR following the Whittaker-Shannon interpolation formula [13]:

$$\tilde{h}(\tau) = \sum_{i=1}^{N} a_i sinc(B_s(t - \tau_i)) \tag{6}$$

where $sinc(\cdot)$ donates to the sinc function defined by $sinc(x) = \frac{sin(\pi x)}{\pi x}$. The sum of normalized sinc functions for every tap in the time-continuous signal allows the reconstruction of the CIR for the band-limited channel.

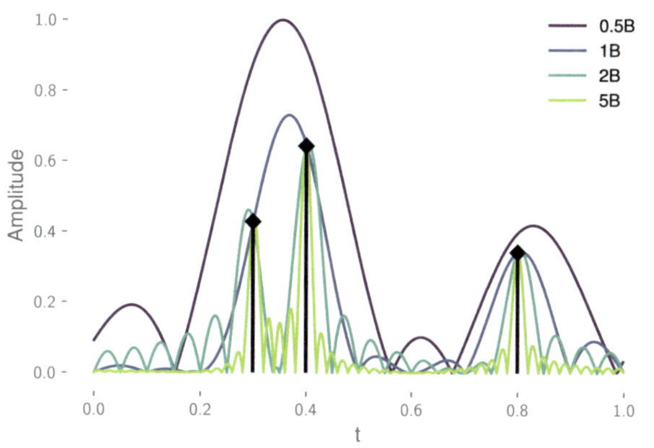

Figure 5.
Qualitative correlation between four different bandwidths and time resolution for bandlimited received signals: MPCs are marked as black Diracs and the signal is modeled as the sum of all sinc function according to Eq. (6). The axes are scaled with min–max scaling, as this is only a qualitative representation of the different bandwidths.

The argument of the sinc function in Eq. (6) is proportional to the used bandwidth. A larger signal bandwidth leads to a narrower sinc function. Thus, more individual MPC in the CIR can potentially be resolved [34]. Therefore, the achievable range resolution Δd for radio sensing is determined by the signal bandwidth B_s [35]:

$$\Delta d = \frac{c}{2B_s} \tag{7}$$

where c donates to the speed of light.

The range resolution is a key metric for many types of radio-sensing tasks and describes the ability to separate different MPC from each other in the CIR. **Figure 5** shows an example of a channel with three multipath components and the bandlimited reconstructed CIRs for different signal bandwidths. For the reconstruction, the signal is interpolated using a sinc kernel according to Eq. (6). The qualitative comparison of different bandwidths shows that a single MPC cannot be resolved if the bandwidth of the signal is not sufficient.

4. UWB radio sensing: Approach and tasks

4.1 Problem formulation

The channel model and impulse response from Eq. (5) can be translated into the spatial domain. The propagation delay τ_i of the ith MPC is the time the electromagnetic wave travels from TX, bouncing at the scatter point (SP) and arriving at the RX. The SP is located somewhere on the target object, which should be located or detected. The time can be converted to a distance d_i by multiplying with the speed of light c [36]:

$$d_i = \tau_i \cdot c = R_{\mathrm{TX,SP}} + R_{\mathrm{SP,RX}} + e_i \tag{8}$$

where $R_{\mathrm{TX,SP}}$ is the geometric distance between TX and SP, and $R_{\mathrm{SP,RX}}$ is the distance between SP and RX. The ranging error is represented by e_i. The measured distance d_i is the length of the propagation path between TX and RX.

For the target localization, we consider a wireless sensor network (WSN) with multiple sensor nodes. Now we can estimate kth different propagation delays $\tau_{i,k}$ for the different channels and also have different propagation path lengths $d_{i,k}$. We assume the WSN consists of an RX at position $\mathbf{X}_{\mathrm{RX}} = [0{,}0{,}0]^{\mathsf{T}}$ and $k = 1, \dots, K$ different TX at positions $\mathbf{X}_{\mathrm{TX},k} = \left[x_{\mathrm{TX},k}, y_{\mathrm{TX},k}, z_{\mathrm{TX},k}\right]^{\mathsf{T}}$. The SP at the target object is located at $\hat{\mathbf{X}}_{\mathrm{SP}} = \left[\hat{x}_{\mathrm{SP}}, \hat{y}_{\mathrm{SP}}, \hat{z}_{\mathrm{SP}}\right]^{\mathsf{T}}$ (**Figure 6**).

$$
\begin{aligned}
\boldsymbol{R}_{\mathrm{bi},k} &= \boldsymbol{R}_{\mathrm{TX}_k,\mathrm{SP}} + \boldsymbol{R}_{\mathrm{SP,RX}} \\
&= \sqrt{\left(x_{\mathrm{TX},k} - \hat{x}_{\mathrm{SP}}\right)^2 + \left(y_{\mathrm{TX},k} - \hat{y}_{\mathrm{SP}}\right)^2 + \left(z_{\mathrm{TX},k} - \hat{z}_{\mathrm{SP}}\right)^2} + \sqrt{\hat{x}_{\mathrm{SP}}^2 + \hat{y}_{\mathrm{SP}}^2 + \hat{z}_{\mathrm{SP}}^2}
\end{aligned} \tag{9}
$$

The bistatic range $\boldsymbol{R}_{\mathrm{bi},k}$ can be obtained or estimated for the measured propagation path $d_{i,k}$ and is then the sum of the transmitter-target range $\boldsymbol{R}_{\mathrm{TX}_k,\mathrm{SP}}$ and the target-receiver range $\boldsymbol{R}_{\mathrm{SP,RX}}$ according to the following equation: [37].

This is a non-linear optimization problem and therefore the solution is not directly obvious. One approach is iterative methods like the Taylor series linearization [38]. Another option is to estimate the target position by a closed-form solution [37, 39] like spherical-interpolation (SI) [40] or spherical-intersection (SX) [41].

4.2 Non-linear least squares estimation

To solve the non-linear equation system given in Eq. (9) the function is linearly approximated at a working point using Taylor's theorem. The solution of the resulting linear least-squares approach is used to adjust the position estimation in an iterative process [38, 42]. A first estimation of the target object position (x_0, y_0, z_0) is utilized to initialize the Taylor series at this point. The innovation of this first estimation $(\delta_x, \delta_y, \delta_z)$ allows the adjustment of the estimation and is calculated as follows:

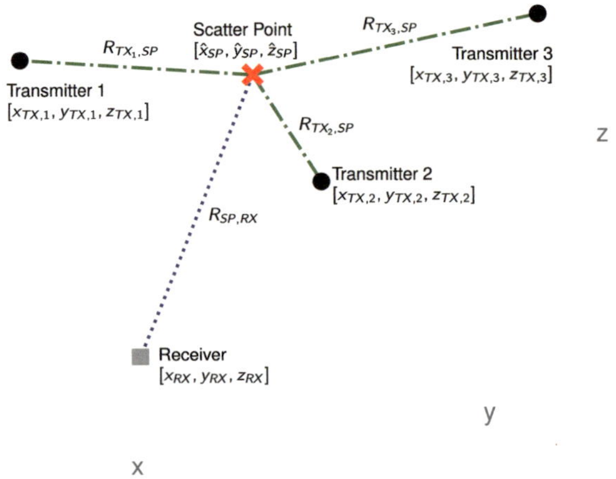

Figure 6.
The geometric configuration of the target and sensors for the localization.

$$\hat{x}_{SP} = x_0 + \delta_x \qquad \hat{y}_{SP} = y_0 + \delta_y \qquad \hat{z}_{SP} = z_0 + \delta_z \tag{10}$$

The first-order Taylor polynomial T_k is used to calculate the linear approximation of Eq. (9) at the first estimation of the target object position:

$$\begin{aligned} T_k &= \boldsymbol{R}_{\text{bi},k} + a_{k,x}\delta_x + a_{k,y}\delta_y + a_{k,z}\delta_z \\ &\approx \boldsymbol{R}_{\text{bi},k}(x_0, y_0, z_0) + \left.\frac{\partial \boldsymbol{R}_{\text{bi},k}(x,y,z)}{\partial x}\right|_{x=x_0,y=y_0,z=z_0}\delta_x + \left.\frac{\partial \boldsymbol{R}_{\text{bi},k}(x,y,z)}{\partial y}\right|_{x=x_0,y=y_0,z=z_0}\delta_y \\ &+ \left.\frac{\partial \boldsymbol{R}_{\text{bi},k}(x,y,z)}{\partial z}\right|_{x=x_0,y=y_0,z=z_0}\delta_z \end{aligned} \tag{11}$$

where $a_{k,x}$, $a_{k,y}$, and $a_{k,z}$ are the partial derivatives of the Eq. (9):

$$\begin{aligned} a_{k,x} &= \frac{x_{\text{TX},k} - x_0}{\sqrt{(x_{\text{TX},k} - x_0)^2 + (y_{\text{TX},k} - y_0)^2 + (z_{\text{TX},k} - z_0)^2}} + \frac{x_0}{\sqrt{x_0^2 + y_0^2 + z_0^2}} \\ a_{k,y} &= \frac{y_{\text{TX},k} - y_0}{\sqrt{(x_{\text{TX},k} - x_0)^2 + (y_{\text{TX},k} - y_0)^2 + (z_{\text{TX},k} - z_0)^2}} + \frac{y_0}{\sqrt{x_0^2 + y_0^2 + z_0^2}} \\ a_{k,z} &= \frac{z_{\text{TX},k} - z_0}{\sqrt{(x_{\text{TX},k} - x_0)^2 + (y_{\text{TX},k} - y_0)^2 + (z_{\text{TX},k} - z_0)^2}} + \frac{z_0}{\sqrt{x_0^2 + y_0^2 + z_0^2}} \end{aligned} \tag{12}$$

In matrix notation this equals:

$$\boldsymbol{A}\boldsymbol{\delta} = \boldsymbol{b}$$

$$\boldsymbol{A} = \begin{bmatrix} a_{1,x} & a_{1,y} & a_{1,z} \\ a_{2,x} & a_{2,y} & a_{2,z} \\ \vdots & \vdots & \vdots \\ a_{k,x} & a_{k,y} & a_{k,z} \end{bmatrix} \qquad \boldsymbol{\delta} = \begin{bmatrix} \delta_x \\ \delta_y \\ \delta_z \end{bmatrix} \qquad \boldsymbol{b} = \begin{bmatrix} d_1 - \boldsymbol{R}_{\text{bi},1} \\ d_2 - \boldsymbol{R}_{\text{bi},2} \\ \vdots \\ d_k - \boldsymbol{R}_{\text{bi,k}} \end{bmatrix} \tag{13}$$

where matrix $\boldsymbol{A}$ represents the geometry matrix or Jacobian matrix containing the partial derivatives for the variables. The vector $\boldsymbol{\delta}$ contains the three-dimensional error components for the object position estimation. Additionally, vector $\boldsymbol{b}$ is the calculated difference between the measured length of the reflection path d_i and the function value at the estimated object position $\boldsymbol{R}_{\text{bi,k}}(x_0, y_0, z_0)$. The linear equation system can then be solved using the least squares approach [38]:

$$\boldsymbol{\delta} = (\boldsymbol{A}^{\intercal}\boldsymbol{A})^{-1}\boldsymbol{A}^{\intercal}\boldsymbol{b} \tag{14}$$

The calculated residual of the position estimation $\boldsymbol{\delta}$ is used to adjust the estimation (Eq. (15)), which is also the starting point for the next iteration of the method:

$$x_0 \leftarrow x_0 - \delta_x, \qquad y_0 \leftarrow y_0 - \delta_y, \qquad z_0 \leftarrow z_0 - \delta_z \tag{15}$$

4.3 Theoretical bounds and geometric dilution of precision

To assess the influence of the geometric constellation between network nodes and scatter points the Cramer-Rao lower bound (CRLB) is computed. The CRLB is the theoretical limit on the performance and accuracy (error variance) of any unbiased estimator and can be derived by the inverse of the Fisher information matrix (FIM). If the FIM is positive definite or non-singular, then the inverse of $\boldsymbol{J}$ exists and the CRLB can be written as [43]:

$$CRLB = \boldsymbol{J}^{-1} \tag{16}$$

The geometric dilution of precision (GDOP) is the ratio of the accuracy limitation of the localization to the accuracy of measurements and is calculated based on the CRLB as follows [44]:

$$GDOP = \sqrt{\operatorname{tr}(CRLB)} = \sqrt{\operatorname{tr}(\boldsymbol{J}^{-1})} \tag{17}$$

where $\operatorname{tr}(\cdot)$ donates to the trace of the square matrix of CRLB. If all measurement errors are to be considered zero-mean independent and identically distributed Gaussian variables in the positioning system the GDOP is [45]:

$$GDOP = \sqrt{\operatorname{tr}\left(\boldsymbol{A}^T\boldsymbol{A}\right)^{-1}} \tag{18}$$

where $\boldsymbol{A}$ represents the Jacobian matrix as in Eq. (12).

The CRLB of the positioning accuracy depends also on the ranging error or error in the MPC extraction [46]. This error also results partly from the limits in the range resolution of UWB (Section 3.4).

Figure 7 presents the calculated GDOP map for every possible target position inside a grid around the sensors for two different sensor arrangements. The GDOP is calculated following Eq. (18) based on the Jacobian matrix $\boldsymbol{A}$ in Eq. (13) for position candidates represented by an equidistant grid. The resulting GDOPs are indicated by the color scale. The first geometric constellation has some areas with degraded GDOP values, while the second symmetrical sensor arrangement around the target results in lower GDOP values and thus a better accuracy [43].

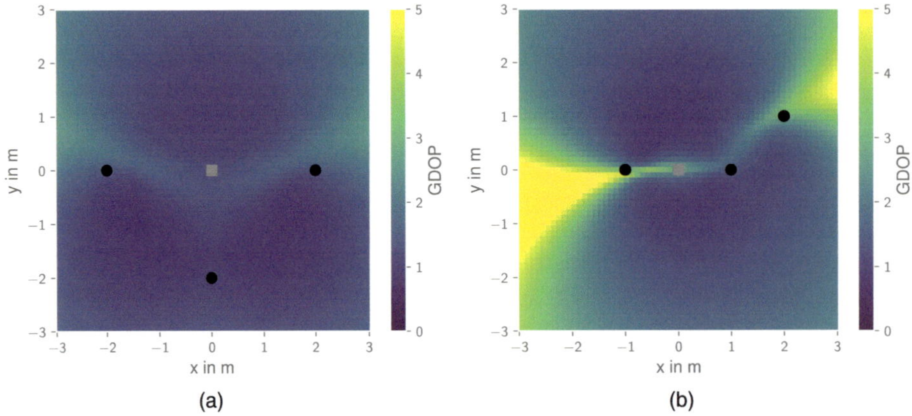

Figure 7.
GDOP map for localization with two different sensor constellations with three TX (black) and one RX (gray). The GDOP (color scale) is calculated for each target position within an equidistant grid sampled with 0.1 m.

4.4 Tasks and functionalities of radio sensing

Radio sensing fulfills different tasks and functionalities depending on the use case (cf. **Table 1**). These tasks include but are not limited to localization, tracking, mapping/imaging, presence detection, counting, or classification and are enabled by specific algorithms. The data processing to achieve the different sensing tasks concerning to UWB specifics is outlined in **Figure 8**. The input for all UWB radio sensing tasks and functionalities is the CIR, from which the clutter is removed beforehand (Section 5.2). Mapping, classification, and counting use all values in the CIR, whereas detection and localization are based on the extracted MPC at the target object.

Figure 9 depicts a collection of results of algorithms and methods to enable radio sensing. All subfigures show original content and were created based on the frameworks and algorithms detailed in [47–49]. The software used to create the illustrations is indicated on the backmatter.

The goal of passive localization is to estimate the position of the target object in a multistatic sensor network based on the measured bistatic ranges between multiple TX and RX. The bistatic range corresponds to the length of the reflection path and can be estimated as MPC from the CIR. An ellipse is defined by the constant bistatic range, on which the target lies, with its foci being located at the TX and RX positions. The target position can either be estimated by setting up the ellipse equation and calculating the intersections of these ellipses or by solving the non-linear equation system (9) with a Taylor series linearization as outlined in Section 4.2. **Figure 9a** shows the localization with the elliptical model with four sensors (three TX and one RX) and the target at the intersection point of the ellipses [42].

Mapping or imaging is accomplished by using the whole CIRs obtained between all TX and the RX. For that, a single CIR is spatially mapped based on the elliptical model resulting in a crowd of ellipses for every distance value in the CIR with the corresponding amplitude values as magnitude. After interpolating the ellipses and combining the resulting grid with the other mapped CIRs in the sensor network, a heatmap of the environment is obtained. The heatmap in **Figure 9b** highlights regions with reflections at the target objects in yellow. The map could also be used to estimate the position of the target objects or even multiple objects [47].

The next task for sensing is the detection of objects or even counting multiple objects/people. Detection is achieved by only analyzing the CIR. As shown in

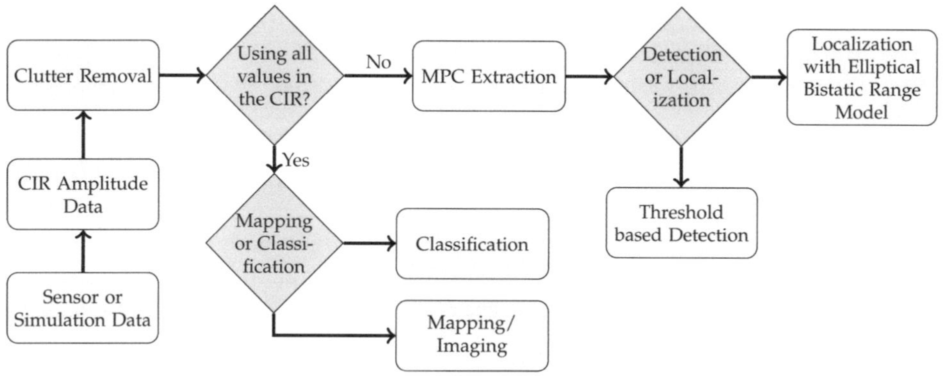

Figure 8.
The connection between data processing and the various sensing tasks.

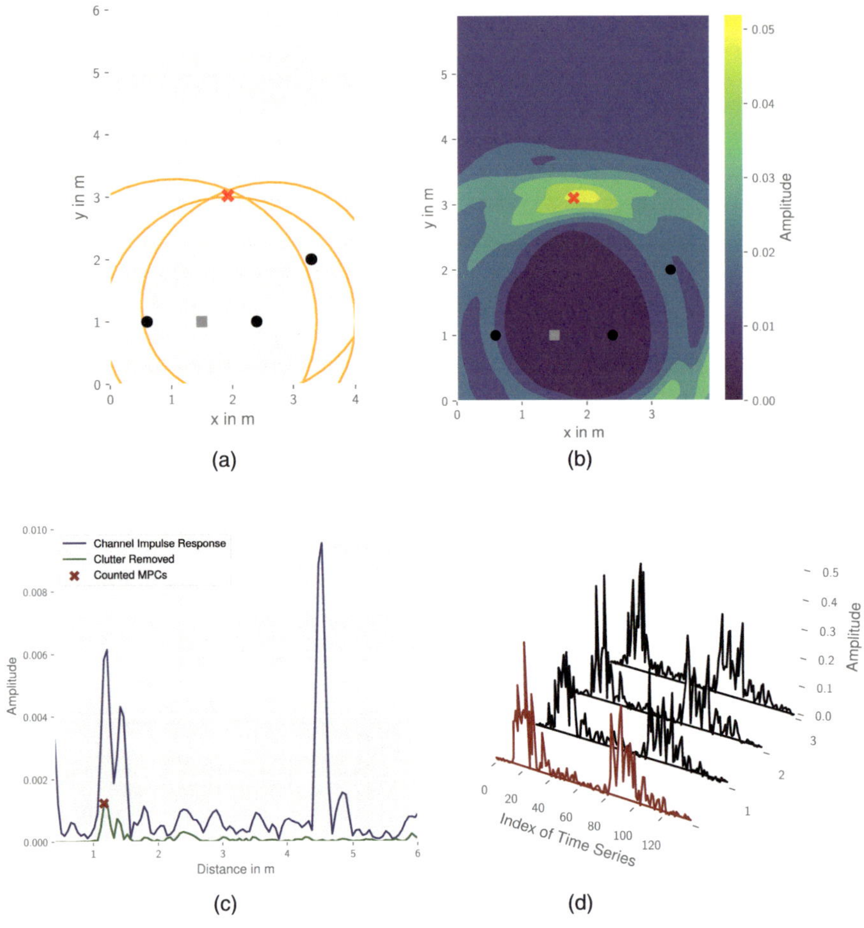

Figure 9.
Tasks of radio sensing: (a) localization with elliptical model, (b) mapping and imaging, (c) detecting and counting with MPC extraction, (d) classification of CIRs with k-nearest neighbor (kNN) algorithm.

Figure 9c, the CIR is filtered to remove the static background using the reference method described in Section 5.2. Then, the MPC from the wanted target is extracted using a simple threshold detector [48]. Further challenges for clutter removal and MPC extraction are discussed in Section 5.2. The CIR and extraction of MPCs could also be used for counting objects or people. Here, the different peaks and local maxima in the CIR are clustered, probability filtered, and the number of targets are extracted by a maximum likelihood estimator [26].

The whole CIR is used for classification for example based on the k-nearest neighbor (kNN) algorithm to distinguish between different states. Therefore, the test data is compared with pre-recorded training data by finding the minimum distance determined by a specific metric. Thus, the recorded CIR could be assigned to a state/class for example to derive a detection status. **Figure 9d** shows an assembled time series from two measured CIRs of the test dataset (red) compared to the three closest time series of the training dataset (black). The kNN algorithm uses a Euclidean metric and the three closest neighbors in the training dataset to determine the state/class of the test sample. The example in **Figure 9d** shows the measured data from seat occupancy detection inside a connected aircraft cabin using UWB. The static background from the CIR is removed beforehand and the CIRs from the different sensors are combined into a unified time series [49].

5. Challenges and research directions

In this section, selected challenges for UWB-based radio sensing are discussed and further research directions are derived. Since the focus in this chapter is on transport and mobility, the challenges are mainly highlighted based on the various use cases in this area.

5.1 Performance indicators and metrics

Moving from solely communication purposes toward perceptive networks fundamentally changes how wireless systems are evaluated. State-of-the-art performance metrics, such as the received signal strength (RSS) or signal-to-interference-plus-noise ratio (SINR), are not useful to evaluate radio sensing systems. Instead, specific metrics for different sensing tasks from other research domains should be applied. For example, vision technologies, such as LiDAR and camera commonly uses the probability of detection [50] to determine the detectability of objects, IPS apply the root mean square error (RMSE) as an accuracy metric, or classification tasks consider the accuracy to predict the label of an object. To further elaborate, **Table 3** proposes and briefly describes performance metrics for different sensing tasks. One challenge is to select the right metric for the desired task and to combine and weigh different metrics.

5.2 Clutter removal and MPC extraction

Clutter is MPCs from the static background environment. In general, these signals are not of interest for sensing applications, but instead only the reflected signals from the target object or detected people need to be considered. The task of clutter removal is to remove or suppress these MPCs in the CIR. One approach is the background subtraction of multipath signals originating from permanent or long-period static objects. There are two methods for background subtraction: the reference method and the dynamic method [5, 7, 51].

The reference method subtracts an averaged reference signal $\overline{h}^{\mathrm{ref}}(t)$ without the target object from the measurement with the target object $h(t)$ flowing Eq. (19). This

Task	Metrics	Description
Localization	Accuracy, mean absolute error (MAE), RMSE	Euclidean distance between estimated position and true position
Tracking	Empirical cumulative distribution functions (ECDF)	MAE or RMSE over the track
Detection	Probability of detection, SINR, prominence, isolation [50]	
Classification	Accuracy, confusion matrix	Confusion matrix to compare predicted and actual class

Table 3.
Possible performance indicators and metrics for different sensing tasks.

method is best suited for static environments and when the calibration with the reference signal is possible beforehand [7, 50]:

$$\overline{h}^{\mathrm{sub}}(t) = |h(t) - \overline{h}^{\mathrm{ref}}(t)| \tag{19}$$

The dynamic method subtracts the static, time-invariant background based on exponential averaging. The background b_t is computed using the previous background estimate b_{t-1} and the newly received CIR h_t [52]:

$$b_t = \alpha b_{t-1} + (1 - \alpha)h_t \tag{20}$$

The constant scalar weighing factor α between 0 and 1 determines whether recent or long-term events are emphasized. The clutter-removed signal $\overline{h}^{\mathrm{sub}}(t)$ is then also obtained by subtraction:

$$\overline{h}^{\mathrm{sub}}(t) = |h_t - b_t| \tag{21}$$

In non-stationary scenarios, clutter removal is much harder, especially in dynamic environments, and can lead to missed detection of the target during the measurement time. Also, background subtraction is not effective in dense multipath scenarios [25].

In addition, the extraction of the MPC with the corresponding time delay from the CIR is of upmost importance to fulfill the different sensing tasks. First, a basic maximum or threshold detector could be used to find the MPC peak in the CIR. More advanced methods cluster the different peaks in the CIR to MPC clusters, which are from a single target object. Other methods used for UWB signals for MPC extraction are detectors of distributed targets (so-called (N,k)-detector), the interperiod-correlation processing (IPCP) detector, and the constant false alarm rate (CFAR) detector [53].

Froehle [36] uses an MPC extraction algorithm consisting of three steps to tackle this challenge. First, all peaks in the CIR are searched with a high-resolution peak search, then a weighting factor is applied before the estimated MPC is detected and canceled out as the strongest scatter. This process is repeated for more attenuated scatter and MPCs [36].

The precondition for the MPC extraction and delay estimation is the distinctness of the MPCs in the CIR. This is especially challenging when the reflected signal strength from a long propagation path is very weak, the target is hidden or shadowed behind other objects, or the MPCs could not be isolated due to limited range resolution and bandwidth [25, 47].

5.3 Channel model and propagation simulation

Commonly the channel for communication systems is modeled with empirical or statistical models like the 3GPP TR 38.901 for 5G [33]. These models are based on measurements of channel statistics for different predefined environment categories such as indoor, urban, or rural. In comparison, deterministic channel models like the multipath channel model described in Section 3 use individual propagation paths or rays as a modeling foundation. This approach is generally suitable for sensing approaches and applications because every ray and MPC can individually be modeled.

Often ray tracing is used to calculate the delay of every propagation path or ray. In addition, the attenuation of the path is determined taking FSPL, reflection, scattering, and diffraction losses into account. Ray tracing is a computationally intensive operation and much more complex than statistical models [54].

An example of a radio sensing mapping task is to estimate the occupancy state of a parking lot in a car park for smart parking and ticketing, as depicted in **Figure 10**. The car park environment is geometrically modeled, and the material properties are applied. Sensors are deployed at different locations (**Figure 10a**), and the signal parameters are complying with UWB [54]. The center frequency is set to 6.5 GHz and the transmit power to -41.3 dBm MHz^{-1}. Then, the individual propagation paths with their corresponding received signal power and propagation delay are computed with the radio propagation simulation software *Altair Feko/WinProp 2022* [55] using the deterministic ray tracing model (**Figure 10a**). After the simulation, the propagation delays and amplitudes are used to obtain the bandlimited reconstructed CIR with help of Eq. (6) in a custom software framework. Since the simulation is performed without and with the target vehicle, the static background subtraction is applied based on the

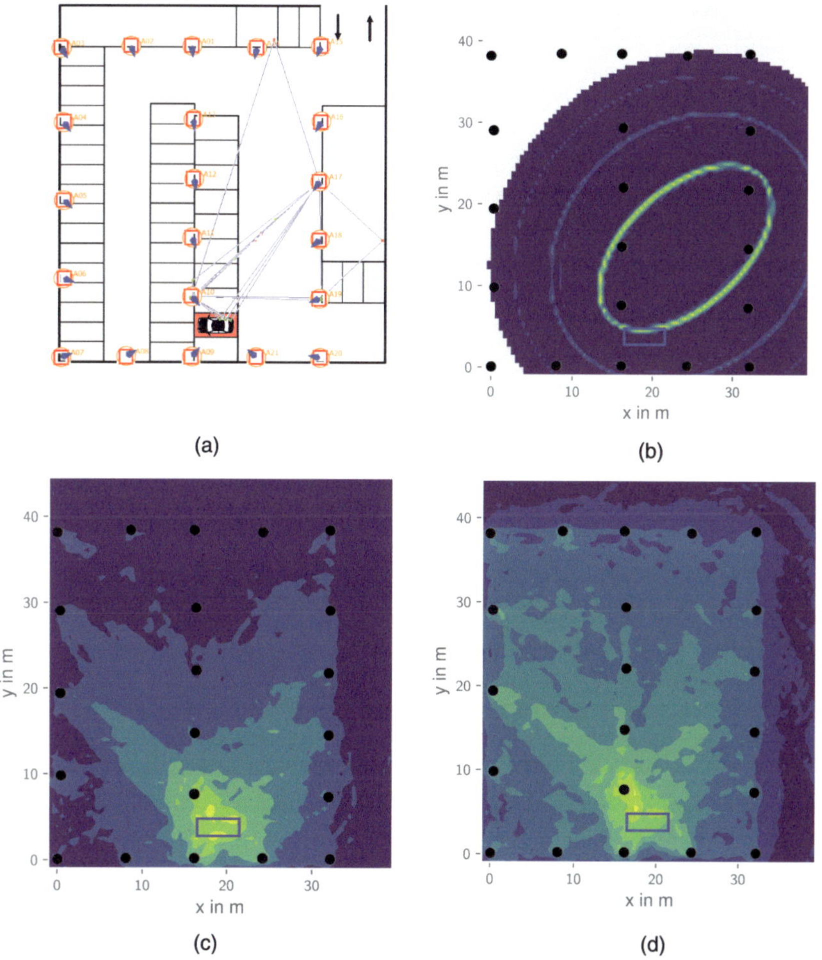

Figure 10.
Antenna comparison for a radio sensing application: (a) radio propagation simulation with Altair Feko/WinProp 2022 [55] for parking plot occupancy detection in a car park with an example of computed rays between two sensors, (b) mapping of a CIR between two sensors (black circles) as a family of ellipses, resulting interpolated heatmap with (c) omnidirectional antennas and (d) directional/sector antennas.

reference method described in Section 5.2. The subtracted CIRs between all sensors are mapped with the elliptical method described in Section 4.4 (**Figure 10b**). Interpolating and combining the maps of all sensor combinations results in a heatmap of the environment highlighting the reflections of the target object for two different antenna configurations (**Figure 10c** and d) [56].

In the future, a combination of deterministic ray tracing and statistical models will be needed to evaluate integrated communication systems with both data transmission and radio sensing capabilities [18].

5.4 Directional antennas

The sensing performance can not only be enhanced with enhanced signal processing but also directly by antennas and the HF signal itself. Omnidirectional antennas radiate the radio power in all directions perpendicular to the azimuth directions equality, whereas specific directional antennas have much higher antenna gain in a specific direction. This means that the signal from this direction is amplified, whereas signals from other directions are much more attenuated. In terms of radio sensing, the reflected signal strength from the desired direction is much stronger in the CIR, so the MPC can be isolated more accurately to find the position of the target. In addition, sector antennas combine multiple directional antenna elements and are switchable to a desired sector and direction. Beamforming even allows spatial selectivity of an antenna array by dynamically controlling the phase and relative amplitude of the signal.

To investigate the influence of directional antennas on the sensing performance, the radio propagation simulation based on ray tracing with *Altair Feko/WinProp 2022* [55] can be used. **Figure 10** compares the mapping/imaging of a car park to estimate the occupancy state of a parking lot with omnidirectional antennas against directional antennas. Due to the selectivity of the antenna and the antenna gain in direction of the target vehicle, reflections are much stronger and highlighted at the boundaries of the vehicle from all sites.

The antenna choice directly affects the radio sensing performance and should be further investigated and considered in addition to algorithms and signal processing improvements.

6. Conclusions

The main contribution of this chapter is the transfer of UWB IR to the emerging research field of radio sensing from a mobility and transportation applications perspective. Therefore, the wideband multipath channel model was introduced and the theoretical bounds for range resolution and GDOP were concluded. Different tasks and functionalities of radio sensing were described across the board ranging from detection, localization, mapping/imaging, counting, and classification. The approaches and algorithms are promising and allow a wide range of use cases of radio sensing in ITS overall modes of transport, for example, smart parking and ticketing, connected aircraft cabin or vehicular networks, and automotive radar.

However, some challenges need to be addressed in further research to implement UWB-based radio sensing systems in real-world applications. These include adaptability to the environment, scalability to a larger sensor network and reliability and robustness of the discussed approaches and algorithms. One approach is machine learning (ML) algorithms to classify the different detection states. This does not require prior MPC extraction. Instead, the entire CIR is used to train and test the classifier with predefined detection scenes and labels to improve the robustness of the detection. Another promising research direction is the integration of radio sensing into a UWB RTLS system to improve the integrity of the localization and enable detection or localization when the object does not carry an active sensor. A fully integrated solution may also use a SLAM approach for anchor mapping, active mobile tag localization, and radio sensing for detection and other tasks.

Software

All calculations, algorithms, and plotting of figures were performed with Python (version 3.10) [57] using the following additional packages: matplotlib (3.6.2) [58], numpy (1.22.4) [59], and scipy (1.9.0) [60]. The radio propagation simulation is carried out with Altair Feko/WinProp 2022 [55].

Author details

Jonas Ninnemann*†, Paul Schwarzbach† and Oliver Michler
Transport Systems Information Technology, Institute of Traffic Telematics, Technische Universität Dresden, Dresden, Germany

*Address all correspondence to: jonas.ninnemann@tu-dresden.de

† These authors contributed equally.

References

[1] Zafari F, Gkelias A, Leung KK. A survey of indoor localization systems and technologies. IEEE Communications Surveys & Tutorials. 2019;**21**(3): 2568-2599. DOI: 10.1109/COMST.2019.2911558

[2] Shi G, Ming Y. Survey of indoor positioning systems based on ultra-wideband (UWB) technology. In: Zeng QA, editor. Wireless Communications, Networking and Applications. Lecture Notes in Electrical Engineering. India: Springer; 2016. pp. 1269-1278. DOI: 10.1007/978-81-322-2580-5_115

[3] Lee G, Park J, Jang J, Jung T, Kim TW. An IR-UWB CMOS transceiver for high-data-rate, low-power, and short-range communication. IEEE Journal of Solid-State Circuits. 2019;**54**(8):2163-2174. DOI: 10.1109/JSSC.2019.2914584

[4] Barneto CB, Turunen M, Liyanaarachchi SD, Anttila L, Brihuega A, Riihonen T, et al. High-accuracy radio sensing in 5G new radio networks: Prospects and self-interference challenge. In: 53rd Asilomar Conference on Signals, Systems, and Computers; 03–06 November 2019. Vol. 2019. Pacific Grove, CA, USA: IEEE; 2019. pp. 1159-1163. DOI: 10.1109/IEEECONF44664.2019.9048786

[5] Zhang JA, Rahman ML, Wu K, Huang X, Guo YJ, Chen S, et al. Enabling joint communication and radar sensing in Mobile networks -a survey. IEEE Communications Surveys Tutorials. 2021;**24**(1):306-345. DOI: 10.1109/COMST.2021.3122519

[6] Chen Y, Zhang J, Feng W, Alouini MS. Radio sensing using 5G signals: Concepts, state of the art, and challenges. IEEE Internet of Things Journal. 2022;**9**(2):1037-1052. DOI: 10.1109/JIOT.2021.3132494

[7] Ledergerber A, D'Andrea R. A multi-static radar network with ultra-wideband radio-equipped devices. Sensors. 2020;**20**(6):1599. DOI: 10.3390/s20061599

[8] Dong J, Guo Q, Liang X. Through-Wall moving target tracking algorithm in multipath using UWB radar. IEEE Geoscience and Remote Sensing Letters. 2021;**19**:3503405. DOI: 10.1109/lgrs.2021.3050501

[9] Le C, Dogaru T, Nguyen L, Ressler MA. Ultrawideband (UWB) radar imaging of building interior: Measurements and predictions. IEEE Transactions on Geoscience and Remote Sensing. 2009;**47**(5):1409-1420. DOI: 10.1109/TGRS.2009.2016653

[10] Choi JH, Kim JE, Kim KT. People counting using IR-UWB radar sensor in a wide area. IEEE Internet of Things Journal. 2021;**8**(7):5806-5821. DOI: 10.1109/JIOT.2020.3032710

[11] Cimdins M, Schmidt SO, Bartmann P, Hellbrück H. Exploiting ultra-Wideband Channel impulse responses for device-free localization. Sensors. 2022;**22**(16):6255. DOI: 10.3390/s22166255

[12] Goldsmith A. Wireless Communications. Cambridge: Cambridge University Press; 2005. pp. 27-98. DOI: 10.1017/CBO9780511841224

[13] Tse D, Viswanath P. Fundamentals of Wireless Communication. Cambridge: Cambridge University Press; 2005. pp. 10-48. DOI: 10.1017/CBO9780511807213

[14] Youssef M, Mah M, Agrawala A. Challenges: Device-free passive localization for wireless environments. In: Proceedings of the 13th Annual ACM International Conference on Mobile Computing and Networking - MobiCom '07. September 2007. Montréal, Québec, Canada: ACM Press; 2007. pp. 222-229. DOI: 10.1145/1287853.1287880

[15] Jovanoska S, Zetik R, Thoma R, Govaers F, Wilds K, Koch W. Device-free indoor localization using a distributed network of autonomous UWB sensor nodes. In: 2013 Workshop on Sensor Data Fusion: Trends, Solutions, Applications (SDF); 09–11 October 2013. Bonn, Germany: IEEE; 2013. pp. 1-6. DOI: 10.1109/SDF.2013.6698264

[16] Mahafza BR, Winton SC, Elsherbeni AZ. Handbook of Radar Signal Analysis. Boca Raton, US: CRC Press; 2021. pp. 93-126

[17] Liu F, Cui Y, Masouros C, Xu J, Han TX, Eldar YC, et al. Integrated sensing and communications: Towards dual-functional wireless networks for 6G and beyond. IEEE Journal on Selected Areas in Communications. 2022;**40**(6): 1728-1767. DOI: 10.1109/ JSAC.2022.3156632

[18] Wild T, Braun V, Viswanathan H. Joint Design of Communication and Sensing for beyond 5G and 6G systems. IEEE Access. 2021;**9**:30845-30857. DOI: 10.1109/access.2021.3059488

[19] Cimdins M, Schmidt SO, Hellbrück H. MAMPI-UWB—Multipath-assisted device-free localization with magnitude and phase information with UWB transceivers. Sensors. 2020;**20**(24):7090. DOI: 10.3390/s20247090

[20] Schmidhammer M, Siebler B, Gentner C, Sand S, Fiebig UC. Bayesian multipath-enhanced device-free localisation: Simulation-and measurement-based evaluation. IET Microwaves, Antennas & Propagation. 2022;**16**(6):327-337. DOI: 10.1049/ mia2.12244

[21] Gentner C, Jost T, Wang W, Zhang S, Dammann A, Fiebig UC. Multipath assisted positioning with simultaneous localization and mapping. IEEE Transactions on Wireless Communications. 2016;**15**(9):6104-6117. DOI: 10.1109/TWC.2016.2578336

[22] Bocus MJ, Piechocki R. A comprehensive ultra-wideband dataset for non-cooperative contextual sensing. Scientific Data. 2022;**9**(1):650. DOI: 10.1038/s41597-022-01776-7

[23] Leitinger E, Meissner P, Rüdisser C, Dumphart G, Witrisal K. Evaluation of position-related information in multipath components for indoor positioning. IEEE Journal on Selected Areas in Communications. 2015;**33**(11): 2313-2328. DOI: 10.1109/ JSAC.2015.2430520

[24] Froehle M, Meissner P, Witrisal K. Tracking of UWB multipath components using probability hypothesis density filters. In: 2012 IEEE International Conference on Ultra-Wideband; 17–20 September 2012. Syracuse, NY, USA: IEEE; 2012. pp. 306-310. DOI: 10.1109/ ICUWB.2012.6340452

[25] Li C, Tanghe E, Fontaine J, Martens L, Romme J, Singh G, et al. Multi-static UWB radar-based passive human tracking using COTS devices. IEEE Antennas and Wireless Propagation Letters. 2022;**21**(4):695-699. DOI: 10.1109/LAWP.2022.3141869

[26] Choi JW, Yim DH, Cho SH. People counting based on an IR-UWB radar sensor. IEEE Sensors Journal. 2017;

17(17):5717-5727. DOI: 10.1109/JSEN.2017.2723766

[27] Qorvo. DW1000 – Qorvo. 2022. Available from: https://www.qorvo.com/products/p/DW1000

[28] Friis HT. A note on a simple transmission formula. Proceedings of the IRE. 1946;**34**(5):254-256. DOI: 10.1109/JRPROC.1946.234568

[29] Kim H. Wireless Communications Systems Design. Chichester, West Sussex. United Kingdom: John Wiley & Sons, Ltd; 2016. pp. 32-51

[30] Pätzold M. Mobile Radio Channels. Chichester, West Sussex. United Kingdom: John Wiley & Sons, Ltd; 2011. pp. 338-342

[31] Hari KVS. Channel models for wireless communication systems. In: Kennington J, Olinick E, Rajan D, editors. Wireless Network Design: Optimization Models and Solution Procedures. International Series in Operations Research & Management Science. New York, NY: Springer; 2011. pp. 47-64. DOI: 10.1007/978-1-4419-6111-2_3

[32] Molisch AF. Ultra-wide-band propagation channels. Proceedings of the IEEE. 2009;**97**(2):353-371. DOI: 10.1109/JPROC.2008.2008836

[33] 3GPP Radio Access Network Working Group. Study on channel model for frequencies from 0.5 to 100 GHz (Release 16). Vol. 16. 3GPP TR 38.901; 2020

[34] Ulmschneider M. Cooperative Multipath Assisted Positioning [Ph.D. thesis]. Hamburg, Germany: Technischen Universität Hamburg; 2021. DOI: 10.15480/882.3299

[35] Wang Z, Han K, Shen X, Yuan W, Liu F. Achieving the performance bounds for sensing and Communications in Perceptive Networks: Optimal bandwidth allocation. IEEE Wireless Communications Letters. September 2022;**11**(9):1835-1839. DOI: 10.1109/LWC.2022.3183235

[36] Froehle M, Meissner P, Gigl T, Witrisal K. Scatterer and virtual source detection for indoor UWB channels. In: 2011 IEEE International Conference on Ultra-Wideband (ICUWB); 14–16 September 2011. Bologna, Italy: IEEE; 2011. pp. 16-20. DOI: 10.1109/ICUWB.2011.6058819

[37] Malanowski M, Kulpa K. Two methods for target localization in multistatic passive radar. IEEE Transactions on Aerospace and Electronic Systems. 2012;**48**(1):572-580. DOI: 10.1109/TAES.2012.6129656

[38] Zaied S. UWB Localization of People - Accuracy Aspects. [Master Thesis]. Ilmenau, Germany: Ilmenau University of Technology; 2009

[39] Noroozi A, Sebt MA. Comparison between range-difference-based and Bistatic-range-based localization in multistatic passive radar. In: 2015 16th International Radar Symposium (IRS); 24–26 June 2015. Dresden, Germany: IEEE; 2015. pp. 1058-1063. DOI: 10.1109/IRS.2015.7226218

[40] Smith J, Abel J. Closed-form least-squares source location estimation from range-difference measurements. IEEE Transactions on Acoustics, Speech, and Signal Processing. 1987;**35**(12):1661-1669. DOI: 10.1109/TASSP.1987.1165089

[41] Mellen G, Pachter M, Raquet J. Closed-form solution for determining emitter location using time difference of arrival measurements. IEEE

Transactions on Aerospace and Electronic Systems. 2003;**39**(3): 1056-1058. DOI: 10.1109/ TAES.2003.1238756

[42] Kocur D, Švecová M, Rovňáková J. Through-The-Wall localization of a moving target by two independent ultra wideband (UWB) radar systems. Sensors. 2013;**13**(9):11969-11997. DOI: 10.3390/s130911969

[43] Godrich H, Haimovich AM, Blum RS. Cramer Rao bound on target localization estimation in MIMO radar systems. In: 2008 42nd Annual Conference on Information Sciences and Systems; 19–21 March 2008. Princeton, NJ, USA: IEEE; 2008. pp. 134-139. DOI: 10.1109/CISS.2008.4558509

[44] Zhang J, Lu J. Analytical evaluation of geometric dilution of precision for three-dimensional angle-of-arrival target localization in wireless sensor networks. International Journal of Distributed Sensor Networks. 2020;**16**(5):1-14. DOI: 10.1177/ 1550147720920471

[45] Lv X, Liu K, Hu P. Geometry influence on GDOP in TOA and AOA positioning systems. In: 2010 Second International Conference on Networks Security, Wireless Communications and Trusted Computing. Vol. 2. Wuhan, China: IEEE; 2010. pp. 58-61. DOI: 10.1109/NSWCTC.2010.150

[46] Jing H, Pinchin J, Hill C, Moore T. An adaptive weighting based on modified DOP for collaborative indoor positioning. The Journal of Navigation. 2016;**69**(2):225-245. DOI: 10.1017/ S037346331500065X

[47] Ninnemann J, Schwarzbach P, Jung A, Michler O. Lab-based evaluation of device-free passive localization using Multipath Channel information. Sensors. 2021;**21**(7):2383. DOI: 10.3390/ s21072383

[48] Ninnemann J, Schwarzbach P, Michler O. Multipath-assisted radio sensing and occupancy detection for smart in-house parking in ITS. In: WiP Proceedings of the Eleventh International Conference on Indoor Positioning and Indoor Navigation - Work-in-Progress Papers (IPIN-WiP 2021); 29 November - 02 December 2021. Lloret de Mar, Spain: CEUR-WS; 2021. pp. 1-15. DOI: 10.48550/ arXiv.2201.06128

[49] Ninnemann J, Schwarzbach P, Schultz M, Michler O. Multipath-assisted radio sensing and state detection for the connected aircraft cabin. Sensors. 2022; **22**(8):2859. DOI: 10.3390/s22082859

[50] Arnold M, Bauhofer M, Mandelli S, Henninger M, Schaich F, Wild T, et al. MaxRay: A raytracing-based integrated sensing and communication framework. In: 2022 2nd IEEE International Symposium on Joint Communications & Sensing (JC&S); 09–10 March 2022. Seefeld, Austria: IEEE; 2022. pp. 1-7. DOI: 10.1109/JCS54387.2022.9743510

[51] Shaikh SH, Saeed K, Chaki N. Moving Object Detection Using Background Subtraction. SpringerBriefs in Computer Science. Cham, Germany: Springer International Publishing; 2014. pp. 15-16. DOI: 10.1007/978-3- 319-07386-6_3

[52] Jovanoska S, Thomä R. Multiple target tracking by a distributed UWB sensor network based on the PHD filter. In: 2012 15th International Conference on Information Fusion; 09–12 July 2012. Singapore: IEEE; 2012. pp. 1095-1102

[53] Rovnakova J, Svecova M, Kocur D, Nguyen TT, Sachs J. Signal processing for through wall moving target tracking by

M-sequence UWB radar. In: 2008 18th International Conference Radioelektronika; 24–25 April 2008. Prague, Czech Republic: IEEE; 2008. pp. 1-4. DOI: 10.1109/RADIOELEK.2008.4542694

[54] Ußler H, Porepp K, Ninnemann J, Schwarzbach P, Michler O. Demo: Deterministic radio propagation simulation for integrated communication Systems in Multimodal Intelligent Transportation Scenarios. In: 2022 IEEE International Conference on Communications Workshops (ICC Workshops); 16–20 May 2022. Seoul, Republic of Korea: IEEE; 2022. pp. 1-2. DOI: 10.1109/ICCWorkshops53468.2022.9915017

[55] Altair Engineering Inc. Altair Feko Applications [Internet]. 2022. Available from: https://www.altair.com/feko-applications [Accessed: April 1, 2023]

[56] Schwarzbach P, Ninnemann J, Michler O. Enabling radio sensing for multimodal intelligent transportation systems: From virtual testing to immersive testbeds. In: 2022 2nd IEEE International Symposium on Joint Communications & Sensing (JC&S); 09–10 March 2022. Seefeld, Austria: IEEE; 2022. pp. 1-6. DOI: 10.1109/JCS54387.2022.9743504

[57] Van Rossum G, Drake FL. Python 3 Reference Manual. Scotts Valley, CA: CreateSpace; 2009

[58] Hunter JD. Matplotlib: A 2D graphics environment computing in science & engineering. Mai. 2007;**9**(3):90-95. DOI: 10.1109/MCSE.2007.55

[59] Harris CR, Millman KJ, van der Walt SJ, Gommers R, Virtanen P, Cournapeau D, et al. Array programming with NumPy. Nature. 2020;**585**(7825):357-362. DOI: 10.1038/s41586-020-2649-2

[60] Virtanen P, Gommers R, Oliphant TE, Haberland M, Reddy T, Cournapeau D, et al. SciPy 1.0: Fundamental algorithms for scientific computing in python. Nature Methods. 2020;**17**(3):261-272. DOI: 10.1038/s41592-019-0686-2